Manual of Water Supply Practices

M49

Quarter Turn Valves: Head Loss, Torque, and Cavitation Analysis

Fourth Edition

AF895852

American Water Works Association

Manual of Water Supply Practices—M49, fourth Edition

Quarter Turn Valves: Head Loss, Torque, and Cavitation Analysis

Disclaimer
The authors, contributors, editors, and publisher do not assume responsibility for the validity of the content or any consequences of their use. In no event will AWWA be liable for direct, indirect, special, incidental, or consequential damages arising out of the use of information presented in this book. In particular, AWWA will not be responsible for any costs, including, but not limited to, those incurred as a result of lost revenue. In no event shall AWWA's liability exceed the amount paid for the purchase of this book.

Sr. Editorial Manager, Books: Suzanne Snyder
Manager – Manuals: Molly Beach
Manager – Publishing Operations: Gillian Wink
Specialist – Copyright and Permissions: Peggy Tyler
Director of Publishing: John Fedor
Technical Editor: Wiley
Cover Composition: Michael Labruyere
Cover Image: wacomka/Shutterstock.com
Production: Innodata

Library of Congress Cataloging-in-Publication Data

Names: American Water Works Association, editor.
Title: Quarter-turn valves : head loss, torque, and cavitation analysis / [edited] by American Water Works Association.
Description: Fourth edition. | Denver, CO : American Water Works Association, [2025] | Series: Manual of water supply practices ; M49 | Includes bibliographical references and index.
Identifiers: LCCN 2024007121 (print) | LCCN 2024007122 (ebook) | ISBN 9781647171575 (paperback) | ISBN 9781613006863 (pdf)
Subjects: LCSH: Water-pipes--Valves.
Classification: LCC TD491 .Q69 2025 (print) | LCC TD491 (ebook) | DDC 628.1/5--dc23/eng/20240315
LC record available at https://lccn.loc.gov/2024007121
LC ebook record available at https://lccn.loc.gov/2024007122

ISBN 978-1-64717-157-5 ISBN, electronic 978-1-61300-686-3

http://doi.org/10.12999/AWWA.M49ed4

American Water Works Association

American Water Works Association
6666 West Quincy Avenue
Denver, CO 80235-3098
303.794.7711
awwa.org

Contents

Figures

Tables

This page intentionally blank.

Preface

The purpose of this manual is to present a recommended method for calculating operating torque, head loss, and cavitation for quarter-turn valves typically used in water works service. It is a discussion of recommended practice, not an American Water Works Association (AWWA) standard. The text provides guidance on generally available methods for using quarter-turn valves as well as their cavitation, flow, and torque characteristics. Questions about specific situations or applicability of specific valves and values should be directed to the manufacturers or suppliers. Information in this manual is useful for technicians and engineers who want a basic understanding of the calculations associated with the use and specification of quarter-turn valves. The valve torque, flow, and cavitation coefficients given are typical but generic values covering a variety of products. Actual flow, cavitation, or torque coefficients for a particular manufacturer's valve should be used in calculations for a specific valve and application to obtain the highest calculation accuracy.

The history of this manual is related to that of American National Standards Institute ANSI/AWWA C504, Standard for Rubber-Seated Butterfly Valves. Until the 1994 edition, ANSI/AWWA C504 included Appendix A, which described a recommended method of calculating torques for butterfly valves. This appendix was deleted from the 1994 and subsequent editions of the standard for several reasons. The AWWA Standards Council directed that standards documents should not contain appendixes; appendix text should either be moved to the main body of the standard or be made into a separate, stand-alone document. Members of the committee for ANSI/AWWA C504 at the time were concerned that the existing text of Appendix A no longer represented the current state of knowledge concerning methods for calculating torques for butterfly valves. In 1993, a subcommittee was established to rewrite Appendix A as a separate manual incorporating the state-of-the-art theory for calculating torque and head loss values for butterfly valves. The second edition of the manual expanded the introduction and some equations, added torque sign conventions, added double offset disc design variables and calculations, added equations for eccentricity torque, added metric units and equivalents, consolidated the nomenclature, and corrected some errors. The third edition manual broadened the application of these methods to include other quarter-turn valves such as ball, plug, and rotary cone valves. This fourth edition provides clarification notes throughout the document and adds Chapter 7 on Control Valve Sizing and Selection.

Manual M49 refers to AWWA standards available for purchase from the AWWA Bookstore. Manufacturers graciously provided valve illustrations and other documentation. AWWA does not endorse any manufacturer's products, and the names of the manufacturers have been removed from the material provided.

This page intentionally blank.

Acknowledgments

The AWWA Standards Council, the AWWA Butterfly Valves Committee, the AWWA Ball and Cone Valves Committee, and the AWWA Eccentric Plug Valves Committee gratefully acknowledge the contributions made by those volunteers who drafted, edited, and provided the significant and critical commentary essential to updating M49. The M49 chapter authors and peer reviewers dedicated many hours in the preparation for this fourth edition to ensure the overall technical quality, consistency, and accuracy of the manual.

Chair

J.R. Holstrom, Val-Matic Valve & Mfg. Corp., Elmhurst, Ill.

AWWA Butterfly Valves Committee (Lead Committee)

Chair
S. Hattan, Tarrant Regional Water District, Fort Worth, Tex.

General Interest Members
M.D. Bennett, Stantec, Cleveland, Ohio
B.E. Bosserman, Consultant, Mission Viejo, Calif.
J. Hebenstreit, Underwriters Laboratories Inc., Northbrook, Ill.
F.L. Hinker, Santa Rosa, N.M.
*J. Hrdlicka,** American Water Works Association, Denver, Colo.
M.C. Johnson, Utah State University, Logan, Utah
J.C. Limke, HDR Inc., Denver, Colo.
J.N. Nakashima, Lockwood, Andrews & Newnam Inc., Oakbrook Terrace, Ill.
*T. Phelan,** AECOM, Kelowna, B.C.
U. Sant, Brown and Caldwell, Rancho Cordova, Calif.
R.A. Ward, Tighe & Bond, Westfield, Mass.

Producer Members
M. Bissauge, Vahn-Tech, Toronto, Ont.
J. Clifton,† McWane Plant & Industrial, Birmingham, Ala.
J.D. Grant,† DeZURIK Inc., Sartell, Minn.
A.C.† *Granzow*, NIBCO, Elkhart, Ind.
T.C. Harbour,† Mueller Water Products, Decatur, Ill.
H. Herold, VAG, Mannheim, Germany
J.R. Holstrom, Val-Matic Valve & Mfg. Corp., Elmhurst, Ill.
K. Johnson, M&H Valve Company, Anniston, Ala.
C. Keeling, IDEX Corporation, Linthicum Heights, Md.
A. Libke, DeZURIK Inc., Royalton, Minn.
T. O'Shea,† Val-Matic, Elmhurst, Ill.
J.H. Wilber, American AVK Company, Littleton, Colo.

User Members
L. Adams, U.S. Bureau of Reclamation, Denver, Colo.
D.W. Coppes, Massachusetts Water Resource Authority, Boston, Mass.
S. Hattan, Tarrant Regional Water District, Fort Worth, Tex.
R. Kilborn, EPCOR, Taylor, Tex.

* Liaison
† Alternate

V.Q. Le, Los Angeles Water and Power, Los Angeles, Calif.
S.E. Metzler,† Trinity River Authority of Texas, Arlington, Tex.
P.J. Ries, Denver Water, Denver, Colo.
S.Y. Tung, City of Houston, Houston, Tex.

AWWA Ball and Cone Valves Committee

Chair
L. Adams, U.S. Bureau of Reclamation, Denver, Colo.

General Interest Members
M.L. Altland, Mott and McDonald, Iselin, N.J.
M.D. Bennett, Stantec, Cleveland, Ohio
B.E. Bosserman, Consultant, Mission Viejo, Calif.
F.L. Hinker, Santa Rosa, N.M.
J. Hrdlicka,* American Water Works Association, Denver, Colo.
C.H. Kirby, Lockwood, Andrews & Newnam Inc., Houston, Tex.

Producer Members
J.D. Grant,† DeZURIK Inc., Sartell, Minn.
H. Herold, VAG, Mannheim, Germany
J.R. Holstrom, Val-Matic Valve & Mfg. Corp., Elmhurst, Ill.
Z.J. Hughes, Ross Valve, Troy, N.Y.
C. Keeling, IDEX Corporation, Linthicum Heights, Md.
A. Libke, DeZURIK Inc., Royalton, Minn.
T. O'Shea,† Val-Matic, Elmhurst, Ill.

User Members
L. Adams, U.S. Bureau of Reclamation, Denver, Colo.
B. Hasanabadi, Colorado Springs Utilities, Colorado Springs, Colo.
S. Hattan, Tarrant Regional Water District, Fort Worth, Tex.
R. Kilborn, EPCOR, Taylor, Tex.
V.Q. Le, Los Angeles Water and Power, Los Angeles, Calif.
P.J. Ries, Denver Water, Denver, Colo.
D. Schwartz,* City of Waynesboro, Waynesboro, Va.
R.E. Tierney, City of Hingham, Hingham, Mass.

AWWA Eccentric Plug Valves Committee

Chair
W.H. Peffley, Springfield, Ill.

General Interest Members
S.D. Adams,† Jacobs, Philomath, Ore.
B.E. Bosserman, Consultant, Mission Viejo, Calif.
F.L. Hinker, Santa Rosa, N.M.
A.R. Holmes, Jones Edmunds, Winter Haven, Fla.
J. Hrdlicka,* American Water Works Association, Denver, Colo.
W.H. Peffley, Springfield, Ill.
J.W. Walker, Kimley Horn, Saint Petersburg, Fla.
Z.R. Yu, Jacobs, Waterloo, Ont.

* Liaison
† Alternate

Producer Members
M.A. Bell, Victaulic, Bainbridge Island, Wash.
D.E. Burczynski,† Kennedy Valve, Elmira, N.Y.
J. Clifton, McWane Plant & Industrial, Birmingham, Ala.
J.D. Grant,† DeZURIK Inc., Sartell, Minn.
J.R. Holstrom,† Val-Matic Valve & Mfg. Corp., Elmhurst, Ill.
A. Libke, DeZURIK Inc., Royalton, Minn.
T. O'Shea, Val-Matic, Elmhurst, Ill.

User Members
S. Hattan, Tarrant Regional Water District, Fort Worth, Tex.
R. Kilborn, EPCOR, Taylor, Tex.
P.J. Ries, Denver Water, Denver, Colo.

AWWA M49 Subcommittee

Chair
J.R. Holstrom, Val-Matic Valve & Mfg. Corp., Elmhurst, Ill.

Members
M.D. Bennett, Stantec, Cleveland, Ohio
W. Biery, Victaulic, Center Valley, Pa.
S. Carpenter, B&K Valves and Equipment, Ramona, Calif.
J. Clifton, McWane Plant & Industrial, Birmingham, Ala.
S. Deboer, Mueller Company, Summerland, B.C.
T.Q. Fallon, Mueller Company, Geneva, Ill.
J.D. Grant, DeZURIK Inc., Sartell, Minn.
T.C. Harbour, Mueller Water Products, Decatur, Ill.
S. Hattan, Tarrant Regional Water District, Fort Worth, Tex.
H. Herold, VAG, Mannheim, Germany
F.L. Hinker, Santa Rosa, N.M.
J.R. Holstrom, Val-Matic Valve & Mfg. Corp., Elmhurst, Ill.
J. Hrdlicka,* American Water Works Association, Denver, Colo.
K. Johnson, M&H Valve Company, Anniston, Ala.
M.C. Johnson, Utah State University, Logan, Utah
R. Kilborn, EPCOR, Taylor, Tex.
A. Le, Emerson Automation Solutions, Englewood, Colo.
A. Libke, DeZURIK Inc., Royalton, Minn.
R.K. Looney, American AVK Company, Minden, Nev.
S.E. Metzler, Trinity River Authority of Texas, Arlington, Tex.
J.N. Nakashima, Lockwood, Andrews & Newnam Inc., Oakbrook Terrace, Ill.
S. Notch, Town of Gilbert, Gilbert, Ariz.
T. O'Shea, Val-Matic, Elmhurst, Ill.
W.H. Peffley, Springfield, Ill.
P.J. Ries, Denver Water, Denver, Colo.
D. Swartzentruber, J&S Valve, Niles, Mich.
J.H. Wilber, American AVK Company, Littleton, Colo.

* Liaison
† Alternate

This page intentionally blank.

Nomenclature

Table 1 provides the variable symbols and definitions used in the manual. Table 2 provides unit conversion between US customary and SI (metric). Table 3 provides abbreviations used in this manual.

Table 1 Nomenclature, terms, and symbols

Term, Variable, or Symbol	Definition or Description	Units US Customary (SI—Metric)
±	Indicates a torque sign convention that may be assigned by the user For valve active torque components: + indicates when the torque tends to close the valve, or – indicates when the torque tends to open the valve. For torque transmitted to the actuator: + indicates when valve shaft torque opposes actuator motion, or – indicates when valve shaft torque assists actuator motion (actuator acts as a brake).	dimensionless
0 though 90	Subscript indicating the valve closure member angle The convention used in this manual is that closed = 0 degrees (0 rad); full open ≈ 90 degrees (π/2 rad).	degrees (rad)
A_{cost}	Annual energy cost	\$/year
A_{pipeid}	Pipe inside flow area	in.^{2} (mm^2)
A_{port}	Valve mean port area based on the center or mean seating point around the perimeter of the seat.	in.^{2} (mm^2)

Term, Variable, or Symbol	Definition or Description	Units US Customary (SI—Metric)
C	Cost of electricity	\$/kWh
C_f	Coefficient of friction between the shaft or trunnion and bushing, dimensionless (This value may be obtained from a flow test, engineering handbooks, the bearing manufacturer, or the valve manufacturer.)	dimensionless
C_g	Valve disc center of gravity distance from shaft centerline	in. (mm)
Closing torque	Test-measured torque in the closing direction	in.-lb (N-m)
$C_{pck\ f}$	Packing force coefficient	lb/in. (N/m)
$C_{pck\ t}$	Packing torque coefficient	lb (N)
C_s	Coefficient of seating, general form used in the first edition	lb/in. (N/m)
C_{sc}	Constant or pressure-independent coefficient of seating torque	lb/in. (N/m)
C_{sp}	Pressure-dependent coefficient of seating torque	lb/in./psi (N/m/kPa)
C_t	Coefficient of dynamic torque (positive value tends to close valve), general form	dimensionless
$C_{t\theta}$	Coefficient of dynamic torque at valve angle θ (positive value tends to close valve)	dimensionless
C_{usc}	Constant or pressure-independent coefficient of unseating torque	lb/in. (N/m)
C_{usp}	Pressure-dependent coefficient of unseating torque	lb/in./psi (N/m/kPa)
C_v	Valve flow coefficient. The flow of water through a valve at 60°F in US gallons per minute at a pressure drop of 1 psi (lb/in.^{2}) Metric Units Note: In metric units, this variable is often identified as K_v. However, this is not used in this manual, as it is easily confused with the resistance coefficient, K. When the resistance coefficient, K, is the resistance coefficient of the valve, it is subscripted with a "V" to indicate this reference, K_v. The metric flow coefficient, K_v, is defined as the flow of water with temperature ranging from 5 to 30°C through a valve in cubic meters per hour (m^3/h) with a pressure drop of 1 bar. (1 bar = 100 kPa). For purposes of this manual, the metric unit version of C_v will be identified by the variable symbol C_{vm}.	gpm/psi^{½} None
C_{vm}	The metric equivalent to C_v (referred to as K_v in other texts)	None (m^3/h/bar^{½}) (m^3/h/[100 kPa]^{½})
D	Nominal valve diameter	in. (mm)

Term, Variable, or Symbol	Definition or Description	Units US Customary (SI—Metric)
D_1	Reducer reduced pipeline diameter	in. (mm)
D_2	Reducer large pipeline diameter	in. (mm)
D_d	Disc diameter for butterfly valves or nominal diameter (D) or port diameter (D_{port}) for ball valves	in. (mm)
D_{pipeid}	Pipe inside diameter	in. (mm)
D_{ref}	Reference valve size	in. (mm)
d_s	Shaft diameter. For trunnion-mounted valves, use trunnion diameter (d_{tr}) in bearing friction calculations.	Inches (mm)
d_{sm}	Mean shaft diameter. Top or actuator shaft and the bottom or blind shaft may be different diameters.	in. (mm)
d_t	Size of model or test valve	in. (mm)
d_{tr}	Trunnion diameter for trunnion-mounted obturators	in. (mm)
E	Efficiency of pump and motor set (80 percent; 0.80, typical)	%
F_f	Liquid critical pressure ratio factor (NOTE: For water in the range of 32°F [0°C] to 125°F [51.7°C], water $F_f \approx 0.96$.)	dimensionless
$F_{l\theta}$ (and $F_{l\theta}^2$)	Liquid pressure recovery factor of a valve without attached fittings. This experimentally determined factor depends upon the type of valve and its internal geometry.	dimensionless
F_r	Resultant force vector for hydrostatic torque	lb (N)
g	Gravitational constant Acceleration due to gravity, 32.2 ft/s^2 (9.81 m/s^2). NOTE: Three or four significant figures are generally practical for most calculations. Critical calculations and raw test data evaluations and reductions should best be done with greater accuracy.	ft/s^2 (m/s^2)
G	Additional seismic acceleration loading multiplier. For applications involving additional seismic loading, component weight may be multiplied by G (horizontal) or G±1 (vertical), where G is the additional seismic acceleration.	Dimensionless
H_p	Packing height	in. (mm)
K	Flow resistance coefficient of any component or fitting	dimensionless
K_1	Reducer resistance coefficient based on the reduced diameter, D_1	dimensionless
K_2	Reducer resistance coefficient based on the large diameter, D_2	dimensionless
K_{sys}	System flow resistance coefficient (excluding the valve)	dimensionless
K_v	Flow resistance coefficient of the valve	dimensionless

Term, Variable, or Symbol	Definition or Description	Units US Customary (SI—Metric)
K_{v90}	Flow resistance coefficient of valve at full open (≈90°, ≈$\pi/2$ rad). NOTE: Use of K_{v90} assumes the valve travels 90 degrees to full open.	Dimensionless
K_{vd1}	Valve resistance coefficient based on the reference diameter, D_1 (typically the nominal diameter, D)	dimensionless
K_{vd2}	Valve resistance coefficient based on the diameter, D_2	dimensionless
$K_{v\theta}$	Flow resistance coefficient of the valve at valve angle θ	dimensionless
L	Reducer end-to-end length	in. (mm)
Opening torque	Test-measured torque in the opening direction	in.-lb (N-m)
P_c	Pressure class or maximum design pressure (the greater of)	dimensionless (or psig)
P_d	Reference downstream pressure for cavitation analysis	psi (kPa)
P_{ew}	Pressure equivalent to disc and shaft weight	psi (kPa)
P_{ml}	Mean seat perimeter length, see Figure 2-11	in. (mm)
PSE	Pressure scale effects factor for cavitation analysis	dimensionless
P_u	Reference upstream pressure for cavitation analysis	psi (kPa)
P_{ua}, $P_{ua\theta}$	Reference pressure just upstream of the valve for choking and cavitation analysis. Subscript θ indicates the valve angle.	psia (kPa)
P_{ut}	Upstream pressure from laboratory test for cavitation analysis	psi (kPa)
P_v	Vapor pressure adjusted for temperature and atmospheric pressure (Example: P_v = –14.4 psig [–99.6 kPa] for water at 60°F [16°C], measured at sea level)	psi (kPa)
P_{va}, $P_{va\theta}$	Absolute vapor pressure (Examples: P_{va} = 0.0867 psia [5.97 kPa] for water at 32°F [0°C] and P_{va} = 1.93 psia [13.3 kPa] for water at 125°F [51.7°C] measured at sea level). Subscript θ indicates the valve angle.	psia (kPa)
P_{vt}	Vapor pressure from laboratory test	psi (kPa)
Q	Volumetric flow rate	gpm (m^3/h)
Q_{max}	Maximum volumetric flow rate	gpm (m^3/h)
R_p	Swing radius of the plug valve's rubber plug face, see Figure 3-9	in. (mm)

Term, Variable, or Symbol	Definition or Description	Units US Customary (SI—Metric)
S_c	Sign convention variable: For torque active components: +1 when the torque tends to close the valve, or –1 when the torque tends to open the valve. For center-of-gravity torque: the sign convention variable is +1 when the center of gravity is above the horizontal and when the disc is in the full-open position, or –1 when the center of gravity is below the horizontal and when disc is fully open. For torque transmitted to the actuator: a positive value when valve shaft torque opposes actuator motion or negative value when valve shaft torque assists actuator motion (actuator is acting as a brake).	dimensionless
S_g	Specific gravity of liquid relative to water at 60°F (16°C) (water = 1.0)	dimensionless
SSE	Size scale effects factor for cavitation analysis	dimensionless
sys	Subscript indicating system piping and components less the butterfly valve	
t	Temperature (water)	°F (°C)
$T_{b0°}$	Bearing torque at valve angle 0 degrees (always positive)	in.-lb (or foot-lb) (N-m)
T_{bt}	Measured bearing torque from testing	in.-lb (N-m)
$T_{btotal\theta}$	Total bearing torque at valve angle θ with addition of disc weight relative to installation orientation (always positive)	in.-lb (or foot-lb) (N-m)
T_{bw}	Bearing torque from disc weight relative to installation orientation (always positive)	in.-lb (or foot-lb) (N-m)
$T_{b\theta}$	Bearing torque at valve angle θ (always positive)	in.-lb (or foot-lb) (N-m)
$T_{cg\theta}$	Center-of-gravity torque at valve angle θ (Positive value tends to close the valve; negative value tends to open the valve.)	in.-lb (or foot-lb) (N-m)
$T_{cg0°}$	Center-of-gravity torque at valve angle 0 degrees (Positive value tends to close the valve; negative value tends to open the valve.)	in.-lb (or foot-lb) (N-m)
$T_{d\theta}$	Dynamic torque at valve angle θ (Positive value tends to close the valve; negative value tends to open valve.) NOTE: Several other names are used in conjunction with this torque component, such as fluidynamic torque, hydrodynamic torque, or aerodynamic torque.	in.-lb (or foot-lb) (N-m)
T_{dt}	Measured dynamic torque from testing (A positive value indicates a tendency to close the valve.)	in.-lb (N-m)
T_{ecc}	Eccentricity torque (Positive value tends to close the valve; negative value tends to open valve.) NOTE: Only considered at the seated position during opening or at closing	in.-lb (or foot-lb) (N-m)

Term, Variable, or Symbol	Definition or Description	Units US Customary (SI—Metric)
T_h	Hydrostatic torque (Positive value tends to close the valve; negative value tends to open the valve.) NOTE: Only considered at the seated position during opening or at closing	in.-lb (or foot-lb) (N-m)
T_p	Packing and hub torque (always positive)	in.-lb (or foot-lb) (N-m)
T_{pt}	Measured packing and hub torque from testing	in.-lb (N-m)
T_s	Seating torque (always positive)	in.-lb (or foot-lb) (N-m)
T_{st}	Measured seating torque from testing (always positive)	in.-lb (N-m)
$T_{t\theta}$	Total operating torque at valve angle θ (Positive value opposes actuator motion; negative value assists actuator motion), general form.	in.-lb (or foot-lb) (N-m)
$T_{tc\theta}$	Total closing torque at valve angle θ (Positive value opposes actuator motion; negative value assists actuator motion.)	in.-lb (or foot-lb) (N-m)
$T_{to\theta}$	Total opening torque at valve angle θ (Positive value opposes actuator motion; negative value assists actuator motion.)	in.-lb (or foot-lb) (N-m)
T_{ts}	Total seating torque (Positive value opposes actuator motion; negative value assists actuator motion.)	in.-lb (or foot-lb) (N-m)
T_{tus}	Total unseating torque (Positive value opposes actuator motion; negative value assists actuator motion.)	in.-lb (or foot-lb) (N-m)
T_{us}	Unseating torque (always positive)	in.-lb (or foot-lb) (N-m)
T_{ust}	Measured unseating torque from testing (always positive)	in.-lb (N-m)
U	Pump usage percentage, 100 percent (1.0) = 24 h per day	%
U_{c1}	Units Conversion Factor: US customary for torque in in.-lb: U_{c1} = 1 in./in. US customary for torque in foot-lb: U_{c1} = 1⁄12 (0.0833) foot/in. Metric for torque in N-m: $U_{c1} = 1 \times 10^{-3}$ (0.001) m/mm	in./in. (foot/in.) (m/mm)
U_{c2}	Units Conversion Factor: US customary for torque in in.-lb: U_{c2} = 1 in./in. US customary for torque in foot-lb: U_{c2} = 1⁄12 (0.0833) foot/in. Metric for torque in N-m: $U_{c2} = 1 \times 10^{-6}$ (0.000001) m^2/mm^2	in./in. (foot/in.) (m^2/mm^2)
V	Velocity of fluid flow	feet per second, fps, or ft/s (meters per second, mps, or m/s)
v	Subscript indicating relative to the quarter-turn valve	
V_{max}	Maximum full-open velocity, fps (NOTE: based on nominal valve diameter if converted from a quantity flow rate)	feet per second, fps, or ft/s (meters per second, mps. or m/s)

Term, Variable, or Symbol	Definition or Description	Units US Customary (SI—Metric)
V_v	Approach fluid velocity of the valve	feet per second, fps, or ft/s (meters per second, mps, or m/s)
V_θ	Approach velocity of fluid flow at valve angle θ	feet per second, fps, or ft/s (meters per second, mps, or m/s)
W_d	Weight of valve disc	lb (kg)
$W_{d\&s}$	Weight of the disc and shaft(s) assembly (banjo). For applications involving additional seismic loading, $W_{d\&s}$ may be multiplied by G or (G±1), where G is the additional gravitational acceleration multiplier.	lb (kg)
W_s	Weight of shaft(s)	lb (kg)
Y	Size scale exponent for cavitation analysis	dimensionless
α	Reducer (increaser) included angle, degrees; for angles < 45 degrees (p/4 radians)	degrees (rad)
β	Pipe angle from vertical axis for center of gravity relative to seat location, 0 to 90 degrees (or 0 to $\pi/2$ rad) when seated position is above horizontal; >90 to 180 degrees (or $\pi/2$ to π rad) when seated position is below horizontal	degrees (rad)
β_r	For reducer flow resistance calculation: Reducer (increaser) diameter ratio	dimensionless
γ	Center-of-gravity offset angle (nonsymmetric disc designs) May also include an adjustment for valve designs where the seating angle is not perpendicular to the pipe axis	degrees (rad)
ΔH	Head loss between any two reference points in a system	feet of water (meters of water)
ΔH_{max}	Head loss across the closed valve or total system with valve closed	feet of water (meters of water)
ΔH_{pipe}	Measured head loss across the pipe during testing without the valve	feet of water (meters of water)
ΔH_{sys}	Head loss across the system	feet of water (meters of water)
ΔH_t	Measured head loss across the valve and pipe during testing	feet of water (meters of water)
ΔH_v	Head loss across the valve, general	feet of water (meters of water)
$\Delta H_{v\theta}$	Head loss across the valve at angle θ	feet of water (meters of water)
ΔP	Pressure drop (or loss) between any two reference points in a system	psid (kPa)
$\Delta P_{dyn\theta}$	Pressure drop (or loss), ΔP_θ, while at closure member angle θ during classic flow or $\Delta P_{choke\theta}$ during choked flow	psid (kPa)

Term, Variable, or Symbol	Definition or Description	Units US Customary (SI—Metric)
$\Delta P_{brg\theta}$	Pressure drop (or loss), ΔP_θ, while at closure member angle θ during classical flow or $P_{ua\theta} - P_{va\theta}$ during choked flow	psid (kPa-d)
ΔP_{choke} and $\Delta P_{choke\theta}$	Differential pressure at onset of choked flow	psid (kPa)
ΔP_{max}	Maximum pressure drop (or loss) across the closed valve or total system with valve closed	psid (kPa)
ΔP_t	Measured pressure drop across the disc from testing	psid (kPa)
ΔP_v	Pressure drop (or loss) across the valve, general form	psid (kPa)
$\Delta P_{v\theta}$	Pressure drop (or loss) across the valve at valve angle θ	psid (kPa)
ΔP_θ	Pressure drop (or loss) while at disc angle θ	psid (kPa)
ε_1	Disc axial offset (NOTE: $\varepsilon_1 = 0$ for symmetric disc designs) See Figure 2-3	in. (mm)
ε_2	Disc lateral offset NOTE: $\varepsilon_2 = 0$ for symmetric or single-offset disc designs Sign Convention NOTE: For hydrostatic torque, ε_2 is positive when oriented above the valve centerline and negative when oriented below the valve centerline. See Figure 2-3.	in. (mm)
θ	Valve opening position angle, closed = 0 degrees (0 rad); full open ≈ 90 degrees ($\pi/2$ rad) Or, as a subscript indicating valve opening position angle for a variable, fully closed = 0 degrees (0 rad); fully open normally equals 90 degrees ($\pi/2$ rad) NOTE: Some designs may not travel the full 90 degrees ($\pi/2$ rad) to the fully open position.	degrees (rad)
θ	Subscript indicating valve opening position angle, fully closed = 0 degrees (0 rad); fully open normally equals 90 degrees ($\pi/2$ rad) NOTE: Some designs may not travel the full 90 degrees ($\pi/2$ rad) to the full-open position.	degrees (rad)
μ_p	Packing coefficient of friction (typically 0.1–0.3)	dimensionless
ν	Packing radial stress to axial stress transfer ratio (typically assumed to be ≈0.5)	dimensionless
ϱ	Fluid (weight) density NOTE: Standard water density is considered as 62.43 lb/ft^3 (1,000 kg/m^3) at 40°F (4.4°C). Also see Table 4. Refer to Crane 2009 (or later). Flow of Fluids Through Valves, Fittings and Pipe, Technical Paper No. 410	lb/ft^3 (kg/m^3)
σ	Cavitation index, general form	dimensionless
σ_c	Constant cavitation index at a reference pressure, P_u	dimensionless
σ_{ct}	Constant cavitation index from laboratory testing	dimensionless
σ_i	Incipient cavitation index at a reference pressure, P_u	dimensionless
σ_{it}	Incipient cavitation index from laboratory test	dimensionless

Term, Variable, or Symbol	Definition or Description	Units US Customary (SI—Metric)
φ	Valve installed shaft angle from vertical axis, 0–90 degrees (or 0 to $\pi/2$ rad)	degrees (rad)
Ω	Pipe angle from vertical axis for hydrostatic and bearing torque, 0–90 degrees (or 0 to $\pi/2$ rad) for flow downhill; 90–180 degrees (or $\pi/2$ to π rad) for flow uphill	degrees (rad)
W_{seat}	Seat width	in. (mm)
K	Seating compression coefficient, typically equal to 2	none
μ_{seat}	Seat friction coefficient	none

Table 2 Conversion of units

US Customary	=	SI (Metric)		SI (Metric)	=	US Customary
1 ft	=	0.3048 m		1 m	=	3.28084 ft
1 ft-lb	=	1.355818 N-m		1 N-m	=	0.737562 ft-lb
1 gallon (liquid)	=	3.785412×10^{-3} m^3		1 m^3	=	264.172037 gallons (liquid)
1 gallon (liquid)	=	3.785412 liters		1 liters	=	0.264172 gallons (liquid)
1 gpm (liquid)	=	0.227124 m^3/h		1 m^3/h	=	4.402867 gpm (liquid)
1 in.	=	25.4 mm		1 mm	=	0.03937 in.
1 in.	=	0.0254 m		1 m	=	39.370078 in.
1 in.-lb	=	0.112985 N-m		1 N-m	=	8.850745 in.-lb
1 lb	=	4.448222 N		1 N	=	0.224809 lb
1 lb	=	0.4535924 kg		1 kg	=	2.204622 lb
1 lb/ft^3	=	16.01846 kg/m^3		1 kg/m^3	=	0.062428 lb/ft^3
1 lb/in.	=	175.12685 N/m		1 N/m	=	0.00571015 lb/in.
1 lb/in./psi	=	25.4 N/m/kPa		1 N/m/kPa	=	0.03937001 lb/in./psi
1 psi	=	6.894757 kPa		1 kPa	=	0.145038 psi
1 psi	=	0.06894757 bar		1 bar	=	14.5038 psi

Table 3 Abbreviations

Abbreviation	Description
AF	Application factor
ANSI	American National Standards Institute
ASME	American Society of Mechanical Engineers
AST	Actuator sizing torque
BFV	Butterfly valve
BV	Ball valve
ISA	International Society for Automation
MRST	Minimum required shaft torque
PV	Plug valve
RCV	Rotary cone valve

Table 4 Typical water properties

Water Temp.		Water Density		Vapor Pressure					
°F	*°C*	*lbm/ft³*	*kg/m³*	*ft abs.*	*m abs.*	*psia*	*kPa*	*psig*	*psig*
32	0.0	62.42	999.9	0.20	0.061	0.087	0.598	–14.609	–100.728
40	4.4	62.43	1,000.0	0.28	0.085	0.121	0.837	–14.575	–100.488
50	10.0	62.41	999.7	0.41	0.125	0.178	1.225	–14.518	–100.100
60	15.6	62.37	999.1	0.59	0.180	0.256	1.762	–14.440	–99.563
70	21.1	62.30	998.0	0.84	0.256	0.363	2.506	–14.333	–98.820
80	26.7	62.22	996.7	1.17	0.357	0.506	3.486	–14.190	–97.840
90	32.2	62.11	994.9	1.61	0.491	0.694	4.788	–14.002	–96.537
100	37.8	62.00	993.1	2.19	0.668	0.943	6.501	–13.753	–94.824
110	43.3	61.86	990.9	2.95	0.899	1.267	8.738	–13.429	–92.588
120	48.9	61.71	988.5	3.90	1.189	1.671	11.523	–13.025	–89.802
130	54.4	61.55	985.9	5.13	1.564	2.193	15.118	–12.503	–86.207
140	60.0	61.38	983.2	6.67	2.033	2.843	19.602	–11.853	–81.723
150	65.6	61.20	980.3	8.58	2.615	3.647	25.142	–11.050	–76.184
160	71.1	61.00	977.1	10.95	3.338	4.639	31.982	–10.057	–69.344
170	76.7	60.80	973.9	13.83	4.215	5.839	40.261	–8.857	–61.065
180	82.2	60.58	970.4	17.33	5.282	7.291	50.267	–7.405	–51.058
200	93.3	60.12	963.0	26.59	8.105	11.101	76.541	–3.595	–24.784
212	100.0	59.83	958.4	33.90	10.333	14.085	97.113	–0.611	–4.213

Property	Polynomial Equations, t in °F
Vapor Pressure *feet of water (absolute)*	$P_{vp} = 3.042\text{E}{-08} \times t^4 - 5.707\text{E}{-06} \times t^3 + 7.211\text{E}{-04} \times t^2 - 2.908\text{E}{-02} \times t + 5.666\text{E}{-01}$
Vapor Pressure *psia*	$P_{vp} = 1.144\text{E}{-08} \times t^4 - 1.927\text{E}{-06} \times t^3 + 2.469\text{E}{-04} \times t^2 - 9.271\text{E}{-03} \times t + 1.877\text{E}{-01}$
Vapor Pressure *psig*	$P_{vp} = 1.144\text{E}{-08} \times t^4 - 1.927\text{E}{-06} \times t^3 + 2.469\text{E}{-04} \times t^2 - 9.271\text{E}{-03} \times t - 1.451\text{E}{+01}$
Water Density *lbm/ft³*	$\rho = 3.675\text{E}{-14} \times t^6 - 2.157\text{E}{-11} \times t^5 + 3.726\text{E}{-09} \times t^4 + 1.684\text{E}{-07} \times t^3 - 1.858\text{E}{-04} \times t^2 + 1.305\text{E}{-02} \times t + 6.218\text{E}{+01}$

Chapter 1

Introduction

Head loss, torque, and cavitation are important considerations in the selection and sizing of quarter-turn valves in water systems. Quarter-turn valve components must be able to withstand the forces and torques generated during use, and the actuator must drive, hold, and seat the valve. The head loss developed across any valve adds to the energy costs of a pumping system. Cavitation may damage a valve or adjacent piping if not controlled.

The topics in this introductory chapter include an explanation of basic quarter-turn valve design elements and their role in predicting operating head loss, torque, and cavitation. Prior editions limited this manual of standard practice to butterfly valves (BFVs). This edition has been expanded to include information on other quarter-turn valves, including ball valves (BVs), rotary cone valves (RCVs), and eccentric plug valves (PVs). Many of the included illustrations are targeted toward BFVs but are generally applicable to all the valves of this scope.

Head-loss characteristics must be known to predict valve operating torque, and system designers also use these data to size a control valve, calculate pump head requirements, and evaluate the energy costs associated with the head loss across the valve in pumping applications. Valve torque is calculated to size actuators and to provide assurance that the valve components can withstand the internal forces produced by the fluid flow and pressure.

Cavitation is analyzed to avoid undesirable sound and vibration and to prevent damage to the valve and adjacent piping. Cavitation data are determined by flow testing. Values for the range of valve angles are helpful in predicting whether cavitation will occur in a given application.

Head loss, torque, and cavitation vary with a valve's position (percentage or angle of opening). These characteristics also depend on the geometry of the valve body and closure member as well as the characteristics of the system in which the valve is installed. Flow testing procedures of a valve comprise a smooth, undisturbed flow upstream and downstream of the valve as produced by a run of straight, constant-diameter pipe. Although variation from this ideal condition influences valve head loss, torque, and cavitation, these conditions are the benchmark and basis for analysis. Flow disturbances caused by piping configuration such as elbows, reducers, or other valves within a distance less than eight times the diameter upstream of the valve may require further review by applying the recommendations given in Chapter 6.

Coefficients provided by the quarter-turn valve manufacturer may be used to calculate the head loss and torque as described in this manual of standard practice, using data based on the test methods described in Chapter 5. The typical coefficients provided in this manual are presented only for illustrative and approximation purposes. Typical information will assist in determining the applicability or sensitivity of some characteristics, in valve type selection, and in many system design considerations.

The closure members of this manual of standard practice are typically referred to as the ball, disc (BFV), cone, or plug. This manual of standard practice may refer to a general closure member or to one specific design. International and European standards will also use the term obturator for the closure member.

SCOPE

The fluid flow and torque calculations are based on water or wastewater flow and do not specifically relate to other liquids or gases. The adjustments for application to other fluids can be found in other texts on fluid mechanics. This manual of standard practice covers round or circular BVs and BFVs within the scopes of AWWA and American National Standards Institute (ANSI) standards ANSI/AWWA C504, Rubber-Seated Butterfly Valves; ANSI/AWWA C507, Ball Valves, 4 In. Through 60 In. (150 mm Through 1,500 mm); ANSI/AWWA C516, Large-Diameter Rubber-Seated Butterfly Valves, Sizes 78 In. (2,000 mm) and Larger; and ANSI/AWWA C519, High-Performance Waterworks Butterfly Valves—3 In. (75 mm) Through 60 In. (1,500 mm), with essentially full-ported designs in which the port diameter and closure member diameter are close to the nominal pipe size (NPS) or nominal diameter (in inches or millimeters). This includes BFVs in sizes 3 in. (75 mm) and larger and BVs in sizes 6 in. (150 mm) through 60 in. (1,500 mm).

This manual of standard practice also covers PVs that have round or oblong-shaped ports and are available with either full or reduced port areas within the scope of ANSI/AWWA C517, Resilient-Seated Cast-Iron Eccentric Plug Valves. Reduced port areas are generally greater than 75 percent of full pipe area.

RCVs are described in AWWA C522, Rotary Cone Valves, 6 In. Through 60 In. (150 mm Through 1,500 mm), in pressure ratings of 150, 250, and 300 for cast- or ductile-iron and steel construction. This type of valve is also included in this manual of standard practice.

Some manufacturers produce valves that are configured as three-way and/or four-way, which have three or four connection ports and require special considerations not included in this manual of standard practice. The valves covered are of the two-way (two end connections, on-off or throttling) configuration. For all of these valves, one must use the matching data for the valve design of interest.

MRST AND AST DEFINITIONS

For purposes of clarity and understanding, many of the AWWA quarter-turn valve standards refer to the operating torque requirements of the valve as two different terms. These are actuator sizing torque (AST) and minimum required shaft torque (MRST). These definitions appear later in this chapter. These are to be considered not as single values but as a series of values (or curves) that vary with valve position. In some cases, one or two (break and/or break and run) conservative or bounding values may be used throughout the entire valve stroke, but in many cases, values at 10 degrees, 5 degrees, or smaller degree increments of valve travel are necessary. The torque predictions of this manual of standard practice provide the most probable operating torque requirement for a valve when operated under the system conditions analyzed. This total operating torque is referred to as the MRST. Depending on the valve type, actuator standard, or manual of standard practice, and the valve's application (on-off or modulating), the MRST is multiplied by an application factor (AF) to obtain an AST (AST = MRST × AF). This is also calculated at many valve positions to correctly size the actuator. See the valve or actuator standards for the AFs to be used.

The actuator sizing additional torque margin, allowances for in-service degradation, and/or safety factors for power (i.e., electric motor, cylinder, or vane) actuators are provided in other ANSI/AWWA standards and included in the AFs and other sizing requirements of the product standard.

DIAMETER ASSUMPTIONS

For the valve shaft diameter, valves meeting ANSI/AWWA C504 have the minimum shaft diameters given in the standard. ANSI/AWWA C516, ANSI/AWWA C507, ANSI/AWWA C517, and ANSI/AWWA C519 do not provide minimum shaft diameters. It is always best to obtain the shaft diameter by measurement or from the manufacturer's documentation.

Many sources are available for quarter-turn valve flow and torque coefficients. These include valve engineering handbooks, published research papers, and valve supplier manuals, catalogues, or bulletins. The manufacturer generally publishes flow coefficients (i.e., C_v, C_{vm}, or K) for most valves. Some manufacturers consider the torque coefficients (C_t) to be proprietary information and may not publish these data.

Much existing data were developed before published standardization methods, and investigators may have based their calculations on different valve diameter measurements. The major valve diameters include NPS, approach pipe inside diameter (ID), valve port diameter, valve seat diameter, and valve closure member diameter (Figure 1-1). Also, various publications use slight variations of these first-principles equations or use different units of measure. The user is cautioned to evaluate and convert such data to the proper format and units of measure. For instance, some BFV manufacturers provide a dynamic torque coefficient for use in the formula, $T_d = C_1 \times \Delta P$. When equated to the basic formula used here, $T_d = C_t \times D^3 \times \Delta P$, it follows that $C_1 = C_t \times D^3$ or $C_t = C_1/D^3$.

If the data were developed based on a BFV disc diameter and the prediction calculations used the nominal diameter, there will be a larger uncertainty in the results than if the disc diameter were used. This manual of standard practice gives direction on what diameter should be used for standardization, consistency, uncertainty, and/or conservatism purposes. However, for many good engineering reasons, much of the older data do not conform to these guidelines. In many instances, the exact approach pipe ID, valve port diameter, and/or valve closure member diameter are not known at the time the calculation is performed. This forces the designer to assume a conservative diameter with greater uncertainty in the results.

For the valves within the scope of this manual of standard practice, the approach pipe ID, valve port diameter, and valve closure member diameter are almost always equal to or less than the valve's nominal diameter when using US customary dimensions. Therefore, the use of the nominal diameter (NPS) as the diameter in torque prediction calculations will often provide a conservatively high torque value (as the diameter appears in the numerator of the equations). The nominal diameter of the valve may be used in these prediction calculations in lieu of the approach pipe ID, valve port diameter, or valve disc diameter as specified with the understanding that the torque results have a higher

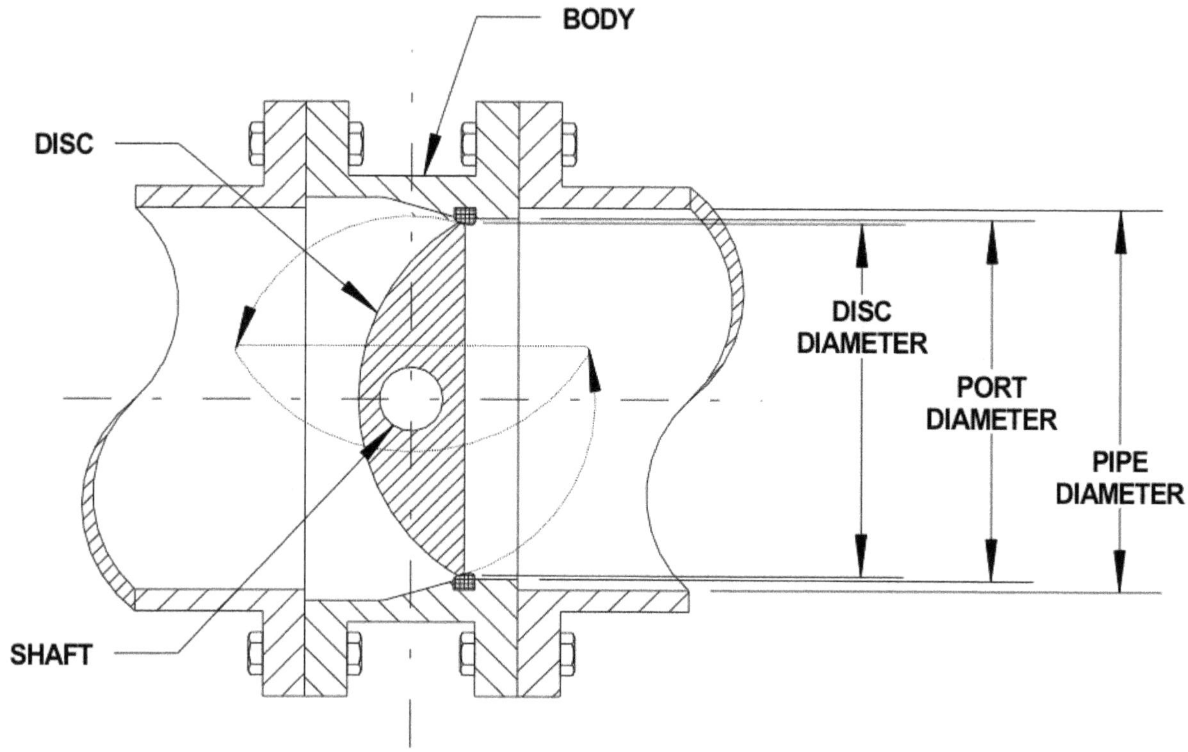

Figure 1-1 Valve disc, port, and pipe diameters

uncertainty and are generally greater than a more precise evaluation. In all cases, if the diameter on which the data are based is known, the use of the same variable provides the highest accuracy prediction.

The flow coefficients, C_v and K, testing and data collection methods that follow are those prescribed in the International Society of Automation (ISA) standard ANSI/ISA S75.02.01, Control Valve Capacity Test Procedures, and are based on the test pipe ID. This methodology does use two slight variations from the ANSI/ISA S75.02.01. First, this practice subtracts the piping loss from the test data to obtain net (valve only) flow coefficient values versus the gross (as measured, including pipe loss) values. Also, the valve shaft axis orientation during the test is vertical. See Chapter 5 for more detail.

QUARTER-TURN VALVE DESIGN

In general, valves may be classified as either linear operation or rotary operation. Linear operating valves include knife gate, slide gate, gate, globe, needle, and diaphragm valves. Rotary operation valves include the BVs, BFVs, cone valves, and PVs in this manual of standard practice. As the full travel of many rotary operation valves approximates a 90-degree rotation, they are often referred to as quarter-turn valves, although travel may be significantly more or less than 90 degrees or a quarter turn. The quarter-turn valve is a versatile component for use as both shutoff and throttling in water systems. Quarter-turn valves are commonly supplied for the water industry in accordance with ANSI/AWWA C504, ANSI/AWWA C507, ANSI/AWWA C516, ANSI/AWWA C519, or ANSI/AWWA C517. As shown in Figures 1-2 through 1-5, these valves consist of a ball, cone, disc, or plug (closure member) supported in the body with a shaft, two stub shafts, or closure member trunnions and bearings. The quarter-turn operation is accomplished with a manual or power actuator connected to a shaft that penetrates the valve body and mounts to the exterior. Valves may have either metallic or elastomeric (rubber or plastic) seats.

Flow is controlled by positioning the closure member between 0 degrees (0 percent, closed) and the full-open (100 percent to approximately 90 degrees) positions. The approximate effective throttling range for quarter-turn valves is 15–75 degrees open (or 15–85 percent), but the range can vary based on application and valve design. Throttling at the lower angles (<15 degrees or 15 percent) may cause erosion due to excessive local velocities or cavitation. Some valves are available with optional throttling or cavitation-reducing trim to extend the control range. See Chapter 4 for discussions of cavitation. Throttling at the higher angles may provide limited control because the valve has little effect on the system flow in many applications.

AWWA Ball Valve Design

AWWA BVs are characterized by the following design elements (Figure 1-2):

- The BV primarily consists of a bored spheroidal closure member (obturator/ball) that rotates roughly a quarter-turn on a shaft or trunnion within the valve body.
- Ports are full nominal size (US customary) and unobstructed (Note: non-AWWA valves may have reduced ports).
- Closure member (ball) may be shaft or trunnion mounted within the body (Note: non-AWWA valves may have a floating ball that is supported by the seats).
- Closure member (ball), when symmetrically mounted, is position seated and does not close tighter by increasing the shaft torque at the seated position.

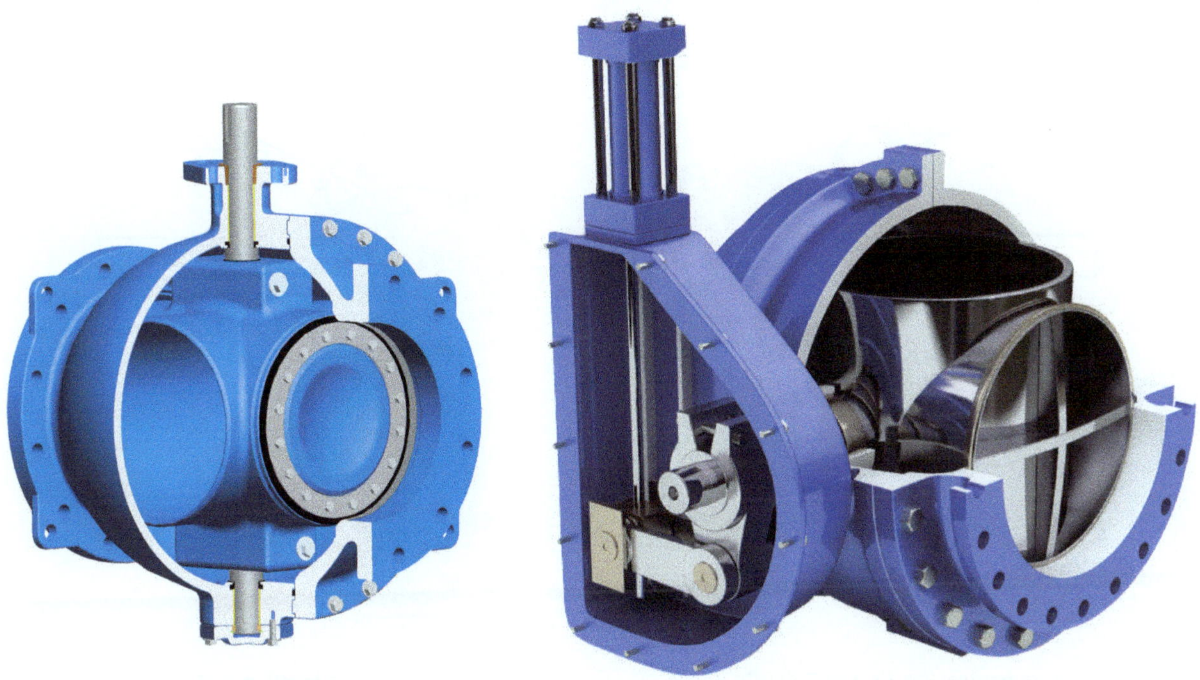

Two-Piece Body Design   Four-Piece Body Design

Source: Val-Matic and DeZURIK

Figure 1-2 Typical BV construction

- Closure member (ball), if eccentrically mounted, may be either position or torque seated. The seat seal will be tightened by increasing the shaft torque against the seat.
- Bodies may be one-, two-, three-, or four-piece construction.
- Seats may be metallic-to-metallic or metallic-to-elastomeric.
- BVs may be single or double seated.
- In double-seated valves, the downstream seat often provides the primary closure seal.
- BVs are often used in pump-control services to control surges and may act as a power-operated check valve.
- In pump-control service, a single-seated BV should be installed to seal tightest against system reverse flow, not pumped flow (this most often places the seat end of the valve toward the pump).
- The BV offers good flow control with a near-equal percentage inherent valve characteristic.
- The full-diameter circular port offers the lowest possible full-open head loss.
- An unobstructed full-open flow path does not produce cavitation or vibration.

AWWA Butterfly Valve Design

AWWA BFVs are characterized by the following design elements (see Figure 1-3):

- The BFV primarily consists of a circular closure member (obturator/disc) that rotates about a quarter turn on a shaft within the valve body.
- These valves have ports that are relatively close to full nominal size of pipe ID.
- Seats may be metallic-to-elastomeric or metallic-to-metallic.

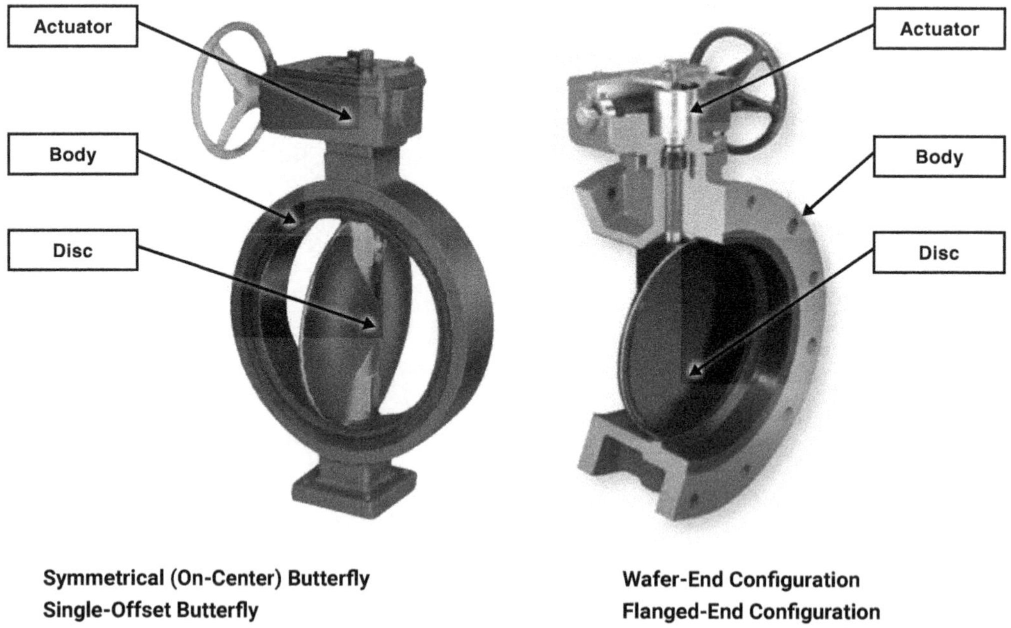

Source: Henry Pratt Co. and DeZURIK

Figure 1-3 Typical BFV construction

- The disc and seats may be symmetric (on-center), single-offset, double-offset, or triple-offset designs.
- Symmetric and single-offset designs are position seated and do not close tighter by increasing the seating shaft torque.
- Double- and triple-offset designs may be actuated as either a position- or torque-seated valve. For torque-seated actuation, the seat's seal will be effectively tightened by increasing the applied shaft torque, which increases the force between the closure member and body seats.
- BFVs may be used in pump-control service to control surges and may act as a power-operated check valve.
- The BFV offers good flow control with a near-equal percentage inherent valve characteristic.
- The BFV offers low full-open head loss.
- Some designs may have a pressure seal direction preference.
- Some designs may have a flow and torque direction preference.

AWWA Plug Valve Design

AWWA PVs are characterized by the following design elements (see Figure 1-4):

- The PV primarily consists of an offset closure member (obturator/plug) that rotates about a quarter turn on a shaft within the valve body.
- These valves have ports that may not be circular and may have either a full or slightly reduced port area.
- Seats are metallic-to-elastomeric.
- The plug and seat are an eccentric design (e.g., double offset).

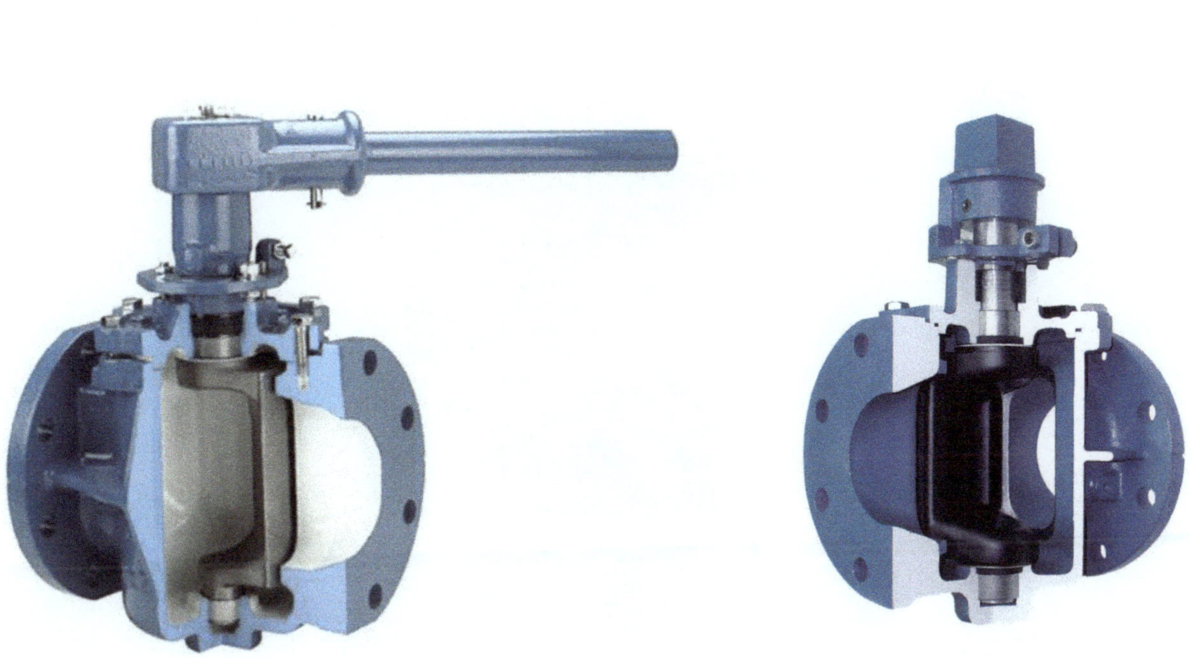

Source: Val-Matic and DeZURIK

Figure 1-4 Typical PV construction

- Materials and construction are designed for both clean water and wastewater service.
- PV may be either position or torque seated, and the seat seal will be tightened by increasing the shaft torque against the seat.
- PVs may be used in pump-control service to control surges and may act as a power-operated check valve.
- In pump-control service, a PV should be installed to seal tightest against system reverse flow, not pumped flow (this generally places the seat end of the valve toward the pump).
- The PV offers good flow control with a near-equal percentage inherent valve characteristic.
- The PV offers moderate to low full-open head loss.
- The normally preferred seal direction installation is the direct pressure orientation.

Rotary Cone Valve Design

RCVs used in the water and wastewater industry are covered by AWWA standard C522 and are typically characterized by the following design elements (see Figure 1-5):

- The RCV primarily consists of a bored and tapered conical closure member (obturator/cone) that rotates a quarter turn on a shaft or trunnion within the valve body and then seals by lowering the tapered cone into the metal seats with an axial movement.
- Actuator provides rotation and torque through the 90-degree operation and thrust or lift at the end of travel positions.
- Ports are full nominal size and unobstructed.
- Body is one piece with a closure bonnet or cover.
- Closure member (cone) is trunnion mounted within the body.

Source: DeZURIK

Figure 1-5 Typical RCV construction

- Closure member (cone) is typically symmetrically (concentrically) mounted and lift seated but does close tighter by increasing the shaft thrust into the seat.
- Seat seals are metallic-to-metallic and generally nickel–copper alloy for severe flow and corrosion-resistant service.
- The seats are tapered and precision ground to mate for drip-tight shutoff.
- Shaft axial movement of the cone into the seat provides the primary closure seal.
- RCVs are often used in pump-control service to control surges and may act as a power-operated check valve.
- The RCV offers good flow control with an equal percentage inherent valve characteristic.
- The full-diameter circular port offers the lowest possible full-open head loss.
- Unobstructed flow path does not produce cavitation or vibration.
- Seat seals are either two- or four-seat configurations.
- Valve can be mounted horizontally or vertically.

Butterfly Valve Offset Designs

BFVs are available in several different offset designs. The discs, shafts, and seats may be aligned in various relative locations. The popular arrangements are the symmetric (on-center) design, the single-offset design, the double-offset design, or the triple-offset design.

In the symmetric (no-offset) design, the disc edge rotates and contacts perpendicularly with a theoretically cylindrical body seat surface (see Figure 1-6). This design is primarily position seated and does not seat tighter with increased applied seating torque. In this configuration, the shaft intersects the disc's seating contact surface.

The single-offset design adds the seat offset (ε_1) and changes the perpendicular cylindrical contact surface into a theoretical conical body seat surface (see Figure 1-6). This design is also primarily position seated and does not seat tighter with increased seating torque. In this configuration and the other offset designs, the shaft does not intersect the disc's seating contact surface.

The double-offset design adds the radial shaft offset (ε_2), which in turn offsets the centerline of the shaft theoretical contact surface cone from the body theoretical contact surface cone, resulting in an eccentric rotation and an effective eccentric wedge angle (see Figure 1-6). This eccentric action and wedge angle cause the seat load to increase with increased applied seating torque, and these valves may be torque seated rather than just position seated.

The triple-offset design rotates the body theoretical contact surface cone of the double-offset design by the wedge angle offset (ε_3) (see Figure 1-6). This increases the eccentric action and wedge angle. Therefore, the seat load increases with increased applied seating torque, and these valves may be torque seated rather than just position seated.

In both the double- and triple-offset designs, the force between the closure member and body seats is controlled by the actuator applied shaft torque. The eccentric rotation of the closure member seat relative to the body seat forces these two seats together for closure or separates them for opening.

Offset valve designs are sensitive to differential pressure or flow direction orientation. Pressure on the shaft side of the disc tends to deflect the disc into the seat, whereas pressure from the seat side of the disc tends to push the disc away from the seat. Because of the lateral offset torque (or eccentricity torque) of the double- and triple-offset designs, the pressure difference in one direction assists seating but opposes unseating. In the opposite orientation, differential pressure opposes seating and assists unseating. Double- and triple-offset valves often have preferred installation orientations typically with the

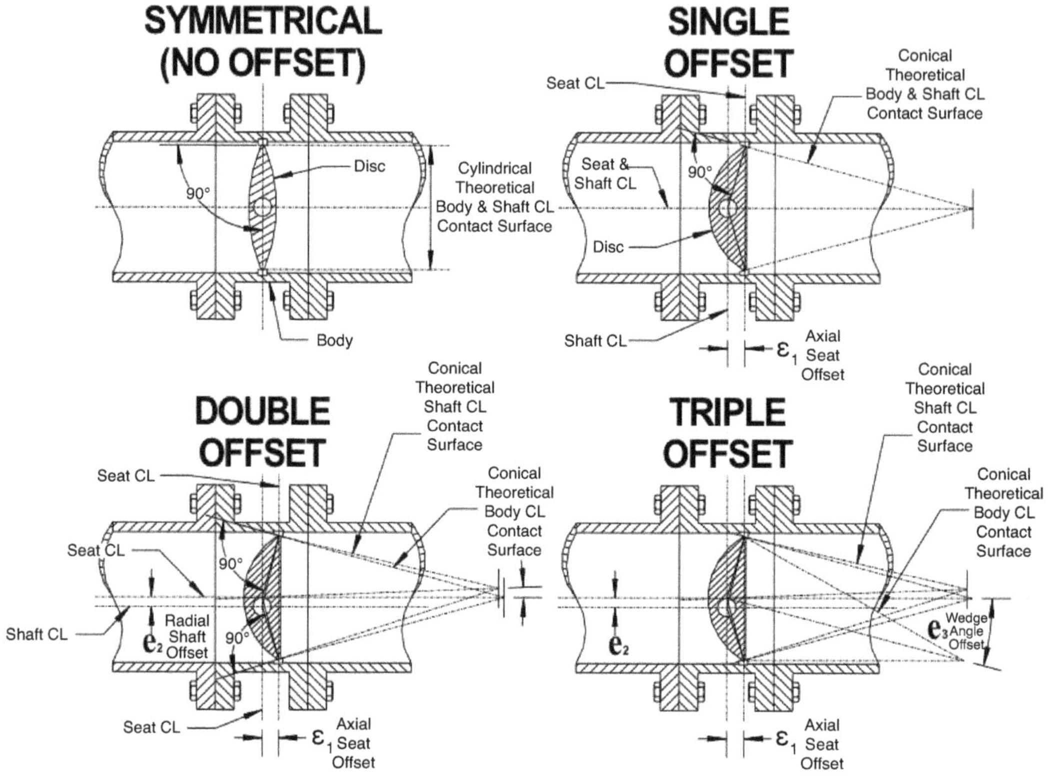

Figure 1-6 BFV offsets

shaft upstream like the "direct" pressure installation of the eccentric PV. However, other torque-, flow-, and cavitation-related issues and concerns might override and require a shaft downstream ("reverse") preferred installation orientation. Always check the flow direction and installation orientation markings on the valve and/or the manufacturer's drawings and operations and maintenance (O&M) documentation before installing.

SYSTEM CONDITIONS

Analysis requires an understanding of system conditions that affect the head-loss, torque, and cavitation calculations for quarter-turn valves, including those conditions in the following list:

1. **Fluid flow rate or velocity:** The maximum anticipated fluid flow rate or flow velocity through the nominal valve size should be determined with consideration of hydraulic design conditions and may include line break or other faulted condition flows when appropriate. The maximum anticipated flow velocity (or flow rate) is needed for the fluid dynamic torque calculations. This is generally at the fully open position of the valve. If not provided at the fully open position, then the valve's differential pressure or the valve's position when this maximum flow rate occurs must also be known and provided for analysis.
2. **Differential pressure:** The maximum (closed) differential pressure is needed for the torque calculations, sealing, and structural considerations. Cavitation calculations also require determination of pressure just upstream and downstream of the valve at the most severe throttling conditions.

3. **Piping installation:** Unless otherwise noted, the valve's inlet and outlet are assumed to be piped conditions. Free discharge outlet (from the valve) and reservoir inlet (to the valve) installations (illustrated in Figure 1-7) represent unique applications that exceed the scope of this manual of standard practice. These installations affect both the head-loss and torque characteristics of a quarter-turn valve. The valve manufacturer should be made aware of these conditions when applicable.
4. **Operating temperature:** Quarter-turn valves and actuators are designed to seat, unseat, control, and rigidly hold the valve closure members under a wide range of operating conditions. Temperature may affect seating torques and friction factors for valve bearings, so it should be considered. The operating temperature of the valves within this scope is 33–125°F (0.6–51.7°C). The valve manufacturer should be consulted or advised when operating temperatures approach or exceed the extremes of this range.
5. **Piping configuration:** Flow turbulence caused by upstream or downstream piping configurations may have a significant effect on valve performance. Nonsymmetrical flow streams or swirling action can magnify the operating torque and head loss of a quarter-turn valve and cause excessive vibration, reducing its useful life. Installation guidelines are presented in Chapter 6.
6. **Maximum upstream line pressure:** The maximum line pressure upstream of the closed valve is not necessary for torque calculation. However, it is needed for cavitation and choking predictions.
7. **Valve location in the system**: The amount of the system that is present upstream and downstream of the valve is used for cavitation and choking evaluations. This can be an estimate of the percentage of the total system that is upstream of the valve.

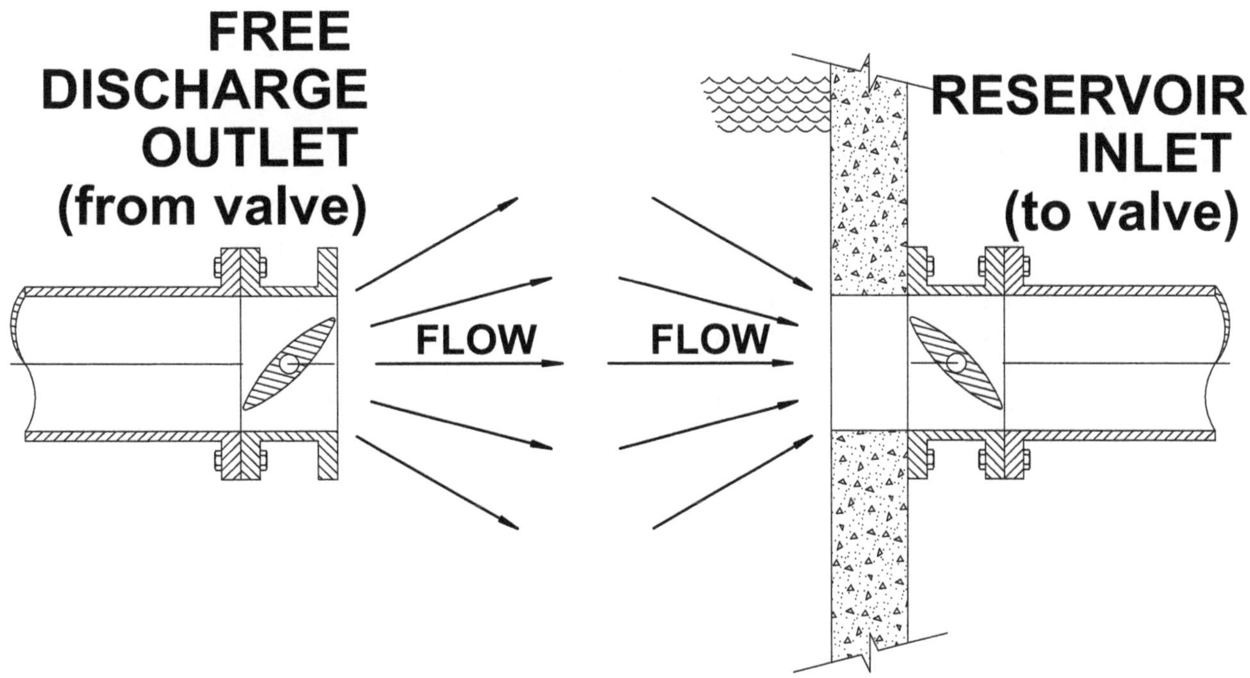

Figure 1-7 Free discharge and reservoir inlet installations of BFVs

DEFINITIONS

The following are definitions used in this manual of standard practice.

Actuator Sizing Torque, AST

The AST is calculated by the valve or actuator manufacturer or the company selecting and mounting the actuator to the valve. The AST is based on the MRST times the AF (AST = MRST × AF). The torque requirements at the seated position (seating or unseating) and the midstroke running positions (5 through 90 degrees, opening or closing) should be evaluated. The AST has different values based on valve position. One or two conservative or bounding values may be used for simplicity, but this is not a single value and should normally be considered as a variable or curve of values.

Application Factor, AF

The AF is a multiplier that is used to determine the actuator's power source sizing. The AF is based on the type of actuator as well as the type of service (on-off service or modulating). The AWWA-recommended AFs are provided in the valve standards.

Closure member

This is a generic term. This may be a ball, disc, knife gate, slide gate, gate, needle, plug, or other term for a specific valve design. Also see the obturator definition.

Direct Pressure

Applied pressure direction in which increasing pressure differential causes seat tightness and friction or seat load to increase as in a single-seated BV or eccentric PV. See Chapter 3 section on "Flow Direction through the Valve" and Figure 3-3.

Equal Percentage Inherent Valve Characteristic

A valve flow characteristic wherein a percentage change in valve position is accompanied by an equal percentage change in the inherent flow coefficient.

Minimum Required Shaft Torque, MRST

The MRST is normally provided by the valve manufacturer and is typically determined by the methods of this manual of standard practice. The torque requirements at the seated position (seating or unseating) and the midstroke running positions (5 through 90 degrees, opening or closing) should be evaluated. The MRST has different values based on valve position. One or two conservative or bounding values may be used for simplicity, but this is not a single value and should be considered as a variable or curve of values. NOTE: For linear valves such as slide gate, gate, and globe, the MRST is often an acronym for minimum required stem thrust (not shaft torque).

Obturator

This is a generic term for the valve's closure member often used in international or European standards. This may be a ball, disc, knife gate, slide gate, gate, needle, plug, or other term for a specific valve design.

Position Convention

The valve position may be designated in degrees open or percentage open. The fully closed position is referred to as the 0 degree or the 0 percent open position. The fully open position is always referred to as the 100 percent open position and normally corresponds to approximately the 90-degree open position. Some quarter-turn valves may stroke more or less than 90 degrees for full travel. This manual of standard practice uses degrees open as the position designation and assumes the 90-degree position and the 100 percent position are equivalent.

Position Seated

A position-seated quarter-turn valve does not seal tighter based on how much torque is applied to the closure member at the seated position. These valves will often permit the closure member to pass completely through the seated position and start to reopen when passing the fully seated position. These valves are typically concentric in their rotation within and through the body seat. These may include symmetric and single-offset valves.

Reverse Pressure

Applied pressure direction in which increasing pressure differential causes seat tightness and friction or load to decrease as in a single-seated BV or eccentric PV. See the Chapter 3 section on "Flow Direction through the Valve" and Figure 3-3.

Seat-Side Flow

This is the fluid flow direction in which the fluid passes the seat centerline before passing the shaft centerline in single-, double-, or triple-offset BFVs; single-seated BVs; and single-seated PVs. See the Chapter 3 section on "Flow Direction through the Valve" and Figure 3-3.

Shaft-Mounted Closure Member

This closure member support design is when a shaft connects to and extends from the closure member directly through a bearing in the valve body. This design has two shaft extensions at diametrically opposed ends of the closure member, and one end penetrates the body for connection to an actuator. The shaft supports the closure member within the valve body and transmits torque for actuation.

Shaft-Side Flow

This is the fluid flow direction in which the fluid passes the shaft centerline before passing the seat centerline in single-, double-, or triple-offset BFVs; single-seated BVs; and single-seated PVs. See the Chapter 3 section on "Flow Direction through the Valve" and Figure 3-3.

Torque Seated

A torque-seated quarter-turn valve seals tighter based on how much torque is applied to the closure member against the seats. These valves will generally not permit the closure member to pass completely through the fully seated position. These valves are typically eccentric in their rotation within and into the body seat. These may include double- and triple-offset BFVs, BVs, and PVs.

Trunnion-Mounted Closure Member

This closure member support design is when a shaft connects to and extends from the closure member directly through the valve body. The body bearings directly support the closure member at diametrically opposed closure member trunnions. This design has one shaft extension that penetrates the body for connection to an actuator and transmitting actuation torque. The shaft does not provide support for the closure member within the valve body.

REFERENCES

ANSI/AWWA. 2023. C504. Rubber-Seated Butterfly Valves. Denver: AWWA.

ANSI/AWWA. 2023. C507. Ball Valves, 4 In. Through 60 In. (150 mm Through 1,500 mm). Denver: AWWA.

ANSI (American National Standard Institute)/AWWA. 2021. C516. Large Diameter Rubber-Seated Butterfly Valves, 78 In. (2,000 mm) and Larger. Denver: AWWA.

ANSI/AWWA. 2021. C517. Resilient-Seated Cast-Iron Eccentric Plug Valves. Denver: AWWA.

ANSI/AWWA. 2018. C519. High-Performance Waterworks Butterfly Valves—3 In. (75 mm) Through 60 In. (1,500 mm). Denver: AWWA.

ANSI/International Society of Automation (ISA). 2012. S75.01.01-2012. Industrial-Process Control Valves - Part 2-1: Flow Capacity - Sizing Equations for Fluid Flow Under Installed Conditions. Research Triangle Park, N.C.: ISA.

ANSI/ISA. 2008. S75.02.01-2008. Control Valve Capacity Test Procedures. Research Triangle Park, N.C.: ISA.

Hutchinson, J.W., ed. 1976 (2nd ed.). *ISA Handbook of Control Valves*. Research Triangle Park, N.C.: ISA.

Chapter 2

Valve Head Loss and Equivalent Resistance System Model

A quarter-turn valve, like any restriction in a pipeline, is a source of head loss. As a quarter-turn valve closes, the head loss increases until the entire system head differential is established across the valve at the closed position. The head loss across a full-open or throttling quarter-turn valve is important because head loss increases energy costs in pumping systems. In control applications, valve head loss is important to determine valve operating positions, control sensitivity, and cavitation potential.

The head loss or pressure drop across a quarter-turn valve can be calculated using many types of flow coefficients. Two commonly used coefficients are discussed in this chapter, and a simple methodology is presented for predicting quarter-turn valve head loss and flow rate during valve operation at midstroke positions.

DISCUSSION OF HEAD LOSS, CHOKING, AND CAVITATION

Most water works analyses can be performed without regard to choking limitations. Choking and cavitation considerations add further calculations and difficulty to the methodology but do not always greatly affect the results of most head-loss, flow, or torque calculations. For more information on choking calculations, also see American National Standards Institute (ANSI)/International Society of Automation (ISA) S75.01.01, Flow Equations for Sizing Control Valves; ANSI/ISA S75.02.01, Control Valve Capacity Test Procedures; and Hutchinson (1976).

The primary calculations of flow rate (or fluid velocity), valve head loss (or pressure drop), dynamic torque, and bearing friction torque given here do not include the effect of choking at the valve. Choking coefficients (such as F_l, F_l^2, or K_m) are required to evaluate choking. The systems where the valves of this scope are used are not normally intended to become fully choked during operation. If choking occurs, additional calculations are required to determine when classic head-loss, flow, and torque equations are no longer appropriate, and additional choked flow calculations should be used. To add choking to the calculations requires system resistance to be determined for the part of the system upstream of the valve as well as the total system equivalent resistance. Analyses including choked flow evaluations are not often performed for the following reasons:

- These calculations add significant complexity to the methodology.
- The line pressure upstream of the closed valve, the differential pressure across the closed valve, and the location of the valve within the system are not known or well-defined. If these inputs are not available, accurate cavitation and choking calculations are not possible.
- In-depth operating system knowledge is not readily available to effectively perform the calculations.
- Actuator sizing and operating torque calculations based on classic head loss and flow only require the maximum closed differential across the valve and the maximum full-open flow rate or fluid velocity.

When choked, the flow rate reaches a maximum limit (relative to the upstream line pressure at the valve), and the dynamic torque does not continue to increase in proportion to increased valve head loss or pressure differential. This is because the flow rate is no longer increasing. Bearing friction torque does continue to increase with increasing valve differential pressure, but the method of determining the differential pressure across the valve changes to the maximum shutoff differential pressure minus the system loss at the choked flow rate or conservatively from upstream pressure to vapor pressure downstream of the valve. The cavitation analysis in this manual can also be used as an indicator of a choking concern. Choking is caused by the development of heavy cavitation resulting in the pressure downstream of the valve reaching vapor pressure.

Figure 2-1 shows typical results of how cavitation and choking occur in a quarter-turn valve application. From a calculation standpoint, the flow and differential pressure are treated as classical flow up to the F_l point and as a constant flow rate at all higher valve differential pressures. There is a calculation error in the transition zone where actual test results are not linear in Figure 2-1. Just like the flow and torque coefficients, the F_l values vary with valve design and position. For critical, highly throttled pressure or flow-control applications, flow calculations that consider cavitation and choked flow at many operating conditions may be necessary to assure good design practices.

DEFINITIONS

Maximum Differential Pressure, ΔP_{max}

The maximum differential pressure (ΔP_{max}) of a quarter-turn valve is defined as the maximum difference between the upstream and downstream pressures when the valve is closed. During operation, the differential pressure varies on the basis of system conditions and valve position. The upstream pressure can be developed from a constant-head source, such as the head from a reservoir or an elevated tank; a variable head source, such as head generated by a pump (see Figure 2-6); or a combination of both. For a conservative torque analysis, the downstream pressure may be assumed to be zero. The differential pressure

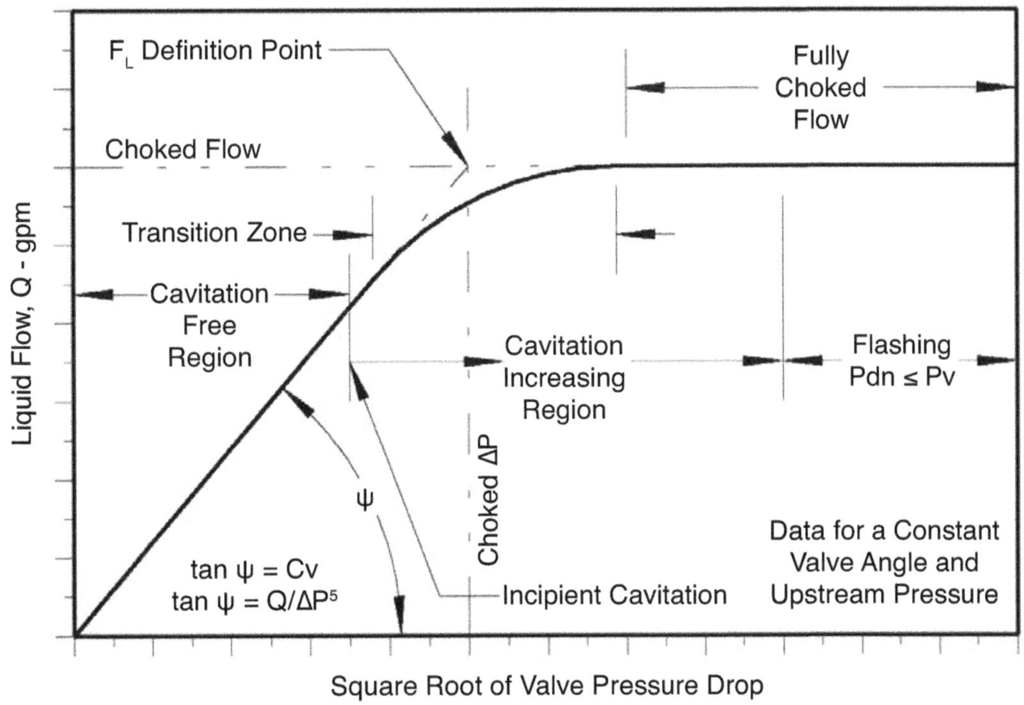

Figure 2-1 Typical quarter-turn valve flow, differential pressure, cavitation, and choking graphical explanation

is used to calculate the forces on the closure member (ball, cone, disc, or plug) and to estimate the flow characteristics of the piping system and calculate the flow rate and valve pressure drop at various valve openings. To determine differential pressure with a variable head created by a pump, the pump curve must be used to calculate the flow through the valve at midstroke valve positions.

Flow Coefficients, General (C_v, C_f, C_d, C_{vm}, K, or K_v)

For any given line velocity or flow rate, a valve's head loss can be predicted by using standard flow equations and flow coefficients. Many flow equations are in use today, designed to satisfy many types of specific operating conditions and flow system applications. There are several variations of flow coefficients, such as K_v, C_v, C_f, and C_d. The two most common flow coefficients used with quarter-turn valves in water service are C_v and K. This manual primarily discusses the use of the resistance coefficient, K or K_v, and flow coefficient, C_v, from the basic fluid equations (Crane 2018). The metric equivalent to C_v is often referred to as K_v in other texts. In this manual, K_v will always refer to the flow resistance coefficient of the valve and not the metric equivalent to C_v. This manual refers to the metric equivalent to C_v as C_{vm} to avoid confusion.

For liquids, the flow coefficient, C_v, expresses the flow capacity in gallons per minute of 60°F (16°C) water with a pressure drop of 1 psi (6.89 kPa). For liquids, the flow resistance (or velocity head loss) coefficient, K, expresses the head loss as directly proportional to the velocity head. The subscript "v" is added to K to indicate when the coefficient is relative to the valve.

Flow coefficients are typically developed using a straight-run test pipe of the same nominal diameter as the valve. The valve may be connected to a pipeline with a slightly different inside diameter (ID). For example, unlined standard-weight 24-in. (600-mm) pipe has an ID of 23.25 in. (590.6 mm). Regardless of true or installed pipe ID, the valve calculations are generally based on nominal valve size, i.e., 24 in. (600 mm). Also, quarter-turn valve inlet diameters or port diameters may be less than the nominal diameters; however, quarter-turn valve torque coefficients are still based on disc or nominal diameter (see Chapter 5). The diameter of the butterfly valve (BFV) disc is usually less than the pipe ID; disc diameter is often used in calculating hydraulic forces on the BFV disc and shaft bearings, but the use of nominal diameter in actuator sizing calculations is a conservative assumption.

Eccentric plug valves (PVs) may have a noncircular port area, but the coefficients are still based on the nominal diameter of the valve. However, the grouping and extension of model test data will also be dependent on the port shape and port area.

Flow Coefficients, C_v and C_{vm}

The C_v valve flow coefficient is often used for control valves (defined previously). Many manufacturers publish C_v values for their valves in the fully open position, as well as the C_v values for intermediate positions or flow characteristic curves, which can be used to determine C_v values in throttling valve positions. These data can be readily used to calculate flow rate or pressure drop in water systems using the equations:

(In US customary units);

$$Q = C_v \times \sqrt{\frac{\Delta P}{S_g}} \quad \text{(Eq 2-1)}$$

(In SI metric units);

$$Q = C_{vm} \times \sqrt{\frac{\Delta P}{S_g}} \quad \text{(Eq 2-2)}$$

$$C_{vm} = C_v \times 0.865 \quad \text{(Eq 2-3)}$$

Where:

Variable	Definition or Description	Units US Customary (SI—metric)
C_v	Valve flow coefficient. The flow of water through a valve at 60°F in US gallon/minute at a pressure drop of 1 psi (lb/in.^{2}) Metric units note: In metric units, this variable is often identified as K_v. However, this is not used in this manual because it is easily confused with the resistance coefficient, K. When the resistance coefficient, K, is the resistance coefficient of the valve, it is subscripted with a "V" to indicate this reference, K_V. The metric flow coefficient, K_v, is defined as: The flow of water with temperature ranging 5–30°C through a valve in cubic meters per hour (m^3/h) with a pressure drop of 1 bar (1 bar = 100 kPa). For the purposes of this manual, the metric unit version of C_v will be identified by the variable symbol C_{vm}.	gpm/psi^{½} (none)
C_{vm}	The metric equivalent to C_v (referred to as K_v in other texts)	none (m^3/h/bar^{½}) (m^3/h/[100 kPa]^{½})
Q	Volumetric flow rate	gpm (m^3/h)

Variable	Definition or Description	Units US Customary (SI—metric)
S_g	Specific gravity of liquid relative to water at 60°F (16°C) (water = 1.0)	dimensionless
ΔP	Pressure drop (or loss) between any two reference points in a system	psid (kPa)

A common misconception is that C_v is sometimes referred to as the "flow capacity" of the valve. Although it is a measure of the valve's capacity to pass water at the specified 1-psi differential pressure, it is not the maximum possible flow rate for the valve or its design flow rate. That is, the flow capacity is not the same as the flow capability. A typical C_v for a 24-in. (600-mm) fully open BFV is about 24,400 gpm/psi^{½}. This statement does not mean that the flow capability of the valve is 24,400 gpm (1.54 m^3/s), which would be equivalent to a flow velocity of 17.3 ft/s (5.3 m/s). BFVs furnished per ANSI/AWWA C504, Rubber-Seated Butterfly Valves, and ANSI/AWWA C516, Large Diameter Rubber-Seated Butterfly Valves, 78 in. (2,000 mm) and Larger, are typically rated for a maximum velocity of 16 ft/s (4.9 m/s). That said, BFVs meeting ANSI/AWWA C504 requirements may also be capable of withstanding and controlling full-open flow velocities well above either 16 ft/s or 17.3 ft/s (4.9 m/s or 5.3 m/s). The C_v (or C_{vm}) is not the same as the valve's design maximum flow rate or its maximum flow capability.

Additionally, to compare the C_v (or C_{vm}) of a valve with other pipe elements such as elbows, tees, or runs of pipe is difficult because these elements are often evaluated as equivalent lengths or a resistance coefficient, K.

Flow Resistance Coefficient, K

The flow resistance coefficient, K, remains relatively constant for various types of valves or fittings through a range of sizes. K is defined as the number of flow velocity heads lost because of a valve or fitting. For example, given that a typical 90-degree elbow has a K coefficient of 0.3 and a tee (branch run) has a K coefficient of 0.9, the system designer can place the 24-in. (600-mm) fully open BFV K of 0.5 (versus a C_v of 24,400) as falling between the normal head losses of an elbow and the branch run of a tee.

The K values for pipes, valves, and fittings in series can be summed directly (for equal-sized components) to find the total flow resistance K of the system (K_{sys}). The K_{sys} can then be used to estimate the flow rate and other parameters of the piping system, using the equivalent resistance model represented in Figure 2-6.

There are some qualifications regarding the applicability of this equivalent resistance methodology.

1. The flow equations and flow resistance coefficients are based on test pipe size. The valve port, disc or closure member, and body diameter may vary from the nominal diameter, as Figure 1-1 illustrates. The user of this method should adjust the coefficients to consider the effects of using a valve in a pipe of larger or smaller ID. This may be done using the beta ratio and relationships provided in Crane (2018) and other fluid mechanics texts. The reader is directed to the Crane (2018) text, as well as Eqs 2-12 through 2-14 in the Reducer Installations section of this chapter.
2. The flow equations are for water service at 60°F (16°C) assuming turbulent incompressible flow. Consult Crane (2009) or similar texts for alternative equations that consider fluid's density, compressible flow, and laminar conditions.
3. Calculations are only as accurate as the test methodology and application used to develop the coefficients. Review laboratory test results whenever possible.

4. Two methodologies are commonly used for testing valves to determine head loss. ANSI/ISA S75.02.01 calls for including the head loss of eight diameters of pipe (two times the pipe diameter upstream and six times the pipe diameter downstream). This manual (and other texts) calls for subtracting the pipe's head loss so that only the valve head loss is reported. When valves with very low head loss are tested, the difference can be significant (40 percent or more). The test procedures given in Chapter 5 are similar to those in ANSI/ISA S75.02.01, with only the valve's net head loss reported. The head loss including the eight diameters of pipe may be referred to as the gross head loss, and the head loss without the pipe loss may be referred to as the net head loss.
5. The accuracy of a calculated head loss is also affected by adjacent piping and fittings. Upstream reducers, elbows, or valves can cause high local velocities, or velocity profile gradients, which may significantly change the valve head loss. Similarly, unusual downstream conditions or free discharge applications produce different results. As the valves of this scope are generally very low-loss valves (e.g., the full-open K_V is generally below 1.0), the ID mismatch between the various types and materials of pipe and fittings, with and without interior linings and coatings, or a gasket that protrudes into the pipe can create significant additional head loss. In the case of the ball valve (BV) or rotary cone valve (RCV), in which the port is full nominal diameter, the full-open resistance is essentially the same as an equal length of the same ID pipe. Any ID changes to or from the upstream and/or downstream pipe or fittings will generate an additional loss that can be greater than the valve itself.

These qualifications clearly suggest that the calculated head loss across a valve should be considered only an estimate, not an exact or calibrated quantity. Its purpose should be limited to energy comparison calculations or overall general system analysis.

Inherent Valve Characteristic

The control valve flow characteristic is the percentage of maximum flow through the valve plotted against travel position. In the case of quarter-turn valves, the travel position may be either percentage of travel or opening angle. The inherent characteristic is that of the valve alone and is the percentage of full-open C_v or C_{vm} plotted against travel position as shown in Figure 2-4.

Installed Valve Characteristic

Once the valve is installed into a system, the flow resistance of the system interacts with that of the valve, and the installed characteristic is that of the valve and system resistance combined as shown in Figure 2-5.

Maximum System Flow Rate or Velocity, Q_{max} or V_{max}

The maximum system flow rate or velocity with the valve fully open is used to calculate valve torques in the range of open positions. If the valve will be operated during temporary high-flow conditions such as fire flows or line-break flows, then the higher flow rate or velocity should be used. Although the maximum system flow rate or velocity is normally assumed to be the same for both the opening and closing operations, to consider extreme or emergency flow rates or velocities for only the operating (opening or closing) stroke of concern may be advantageous. This manual is generally based on the flow expressed as a fluid velocity in the pipe, not as a quantity such as gallons per minute (gpm), liters per minute (L/m), cubic feet per second (ft^3/s), or cubic meters per second (m^3/s). Conversions from quantity to pipe velocity are standard engineering practice and are not discussed in detail. This chapter provides the equivalent resistant system model that determines velocities (or flow rates) and pressures used for the torque and cavitation calculations.

HEAD-LOSS CALCULATIONS

Given the K (or K_V for K of a valve) flow coefficient for a valve or fitting, head loss can be calculated with the following formula from Crane (2018):

$$\Delta H = \frac{K \times V^2}{2 \times g} \qquad \text{(Eq 2-4)}$$

If the flow coefficient is expressed as a C_v, it can be equated to K by:

(In US customary units);
$$K = \frac{891 \times D^4}{C_v^2} \qquad \text{(Eq 2-5)}$$

If the flow coefficient is expressed as a C_{vm}, it can be equated to K by:

(In SI metric units);
$$K = \frac{8.52 \times (D \times U_{c1})^4}{C_{vm}^2} \qquad \text{(Eq 2-6)}$$

Also, if the flow, Q, is expressed in gallons per minute, the fluid velocity can be found by:

(In US customary units);
$$V = \frac{0.4085 \times Q}{D_{pipeid^2}} \qquad \text{(Eq 2-7)}$$

If the flow, Q, is expressed in cubic meters per hour, the fluid velocity can be found by:

(In SI metric units);
$$V = \frac{353.7 \times Q}{D_{pipeid^2}} \qquad \text{(Eq 2-8)}$$

Where:

Variable	Definition or Description	Units US Customary (SI—Metric)
C_v	Valve flow coefficient. The flow of water through a valve at 60°F in US gallons per minute at a pressure drop of 1 psi (lb/in.^{2}) Metric units note: In metric units, this variable is often identified as K_v. However, this is not used in this manual, as it is easily confused with the resistance coefficient, K. When the resistance coefficient, K, is the resistance coefficient of the valve, it is subscripted with a "v" to indicate this reference, K_v. The metric flow coefficient, K_v, is defined as: The flow of water with temperature ranging 5–30°C through a valve in cubic meters per hour (m^3/h) with a pressure drop of 1 bar (1 bar = 100 kPa). For the purposes of this manual, the metric unit version of C_v will be identified by the variable symbol C_{vm}.	gpm/psi^{½} (none)

Variable	Definition or Description	Units US Customary (SI—Metric)
C_{vm}	The metric equivalent to C_v (referred to as K_v in other texts)	none ($m^3/h/bar^{½}$) ($m^3/h/(100\ kPa)^{½}$)
D	Nominal valve diameter	in. (mm)
D_{pipeid}	Pipe ID	in. (mm)
g	Gravitational constant Acceleration due to gravity, 32.2 ft/s² (9.81m/s²). NOTE: Three significant figures are practical for most calculations. Critical calculations and raw test data evaluations and reductions are best performed with greater accuracy.	ft/s^2 (m/s^2)
K	Flow resistance coefficient of any component or fitting	dimensionless
Q	Volumetric flow rate	gpm (m^3/h)
U_{c1}	Units' conversion factor: US customary for torque in in.-lb: U_{c1} = 1 in./in. US customary for torque in ft-lb: U_{c1} = 1⁄12 (0.0833) ft/in. Metric for torque in N-m: $U_{c1} = 1 \times 10^{-3}$ (0.001) m/mm	in./in. (ft/in.) (m/mm)
V	Velocity of fluid flow	feet per second, fps or ft/s, (meters per second, mps, or m/s)
ΔH	Head loss between any two reference points in a system	feet of water (meters of water)

Several other types of flow equations are given in the references cited in the bibliography.

REDUCER INSTALLATIONS

Using a smaller-than-line-size quarter-turn valve for pressure or flow-control applications is common. The pipeline will therefore include pipe reducers on either side of the valve (see Figure 2-2). Reducer nomenclature and geometry are shown in Figure 2-3.

Flow resistance coefficients (K) of elements in series are added directly but must be relative to a constant diameter. In analysis based on the valve's diameter (the smaller diameter), the flow resistance of the reducer and expander sections should be evaluated as K_1 or on the basis of the valve diameter (D or D_1). In this manner, these resistance values can be added to the valve's flow resistance value (K_v) and treated as a unit in the analysis using the diameter of the valve as the conversion of flow rate to velocity. If the analysis is to be based on the large pipeline diameter as the conversion of flow rate to velocity, the reducer K_1, increaser K_1, and valve K_{vd1} should be divided by the beta ratio to the fourth power (β_r^4, Eq 2-12) and added or added together and the sum divided by β_r^4 to obtain the K_2 equivalent.

The resistance coefficient, K_1, for an upstream reducer can be calculated by the following formulas from Crane (2009):

$$\alpha = 2 \times \mathrm{ATAN}\left(\frac{D_2 - D_1}{2 \times L}\right) \qquad \text{(Eq 2-9)}$$

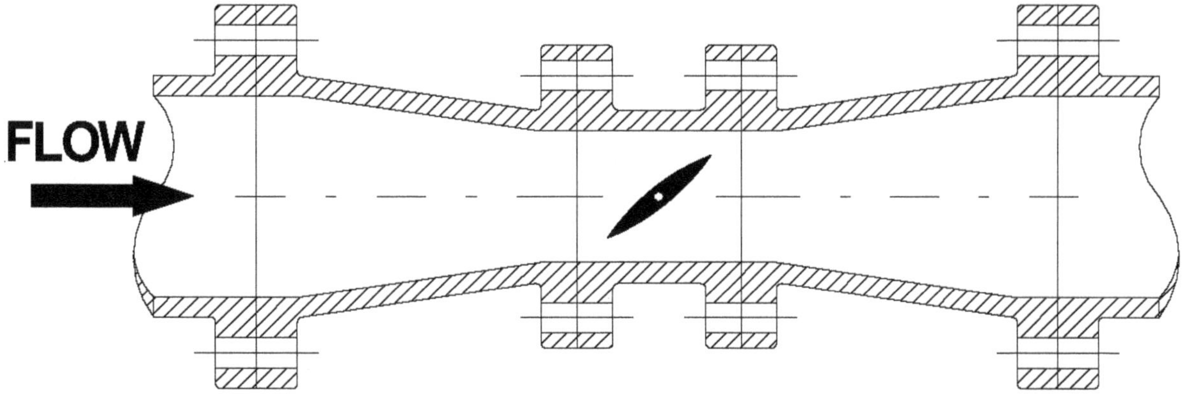

Figure 2-2 Smaller-than-line-size BFV installation

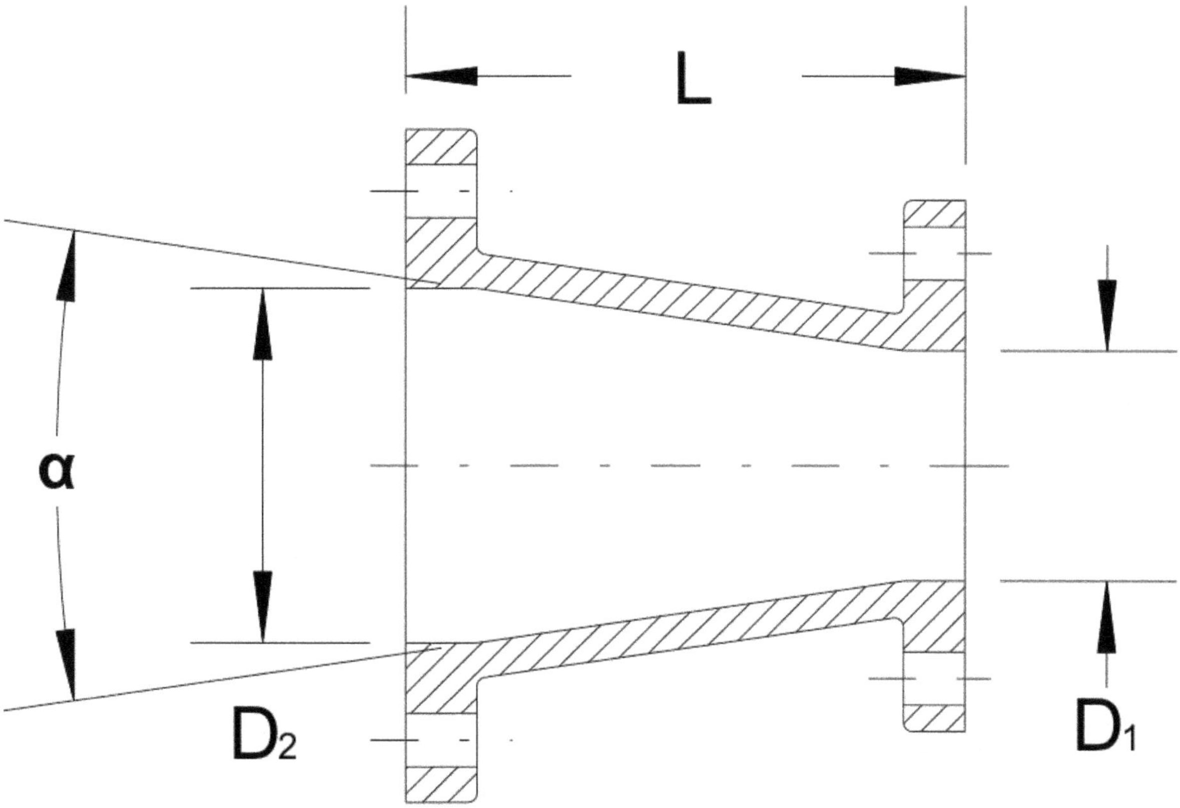

Figure 2-3 Reducer geometry

$$K_1 = 0.8 \times \left(\text{SIN}\frac{\alpha}{2}\right) \times \left[1 - \left(\frac{D_1}{D_2}\right)^2\right] \qquad \text{(Eq 2-10)}$$

The resistance coefficient, K_1, for a downstream reducer (increaser) can be calculated using Eq 2-9 and the following formula:

$$K_1 = 2.6 \times \left(\text{SIN}\frac{\alpha}{2}\right) \times \left[1-\left(\frac{D_1}{D_2}\right)^2\right]^2 \qquad \text{(Eq 2-11)}$$

To base on the pipeline diameter, K_2, use the following equations:

$$\beta_r = \frac{D_1}{D_2} \qquad \text{(Eq 2-12)}$$

$$K_2 = \frac{K_1}{\beta_r^4} \qquad \text{(Eq 2-13)}$$

$$K_{vd2} = \frac{K_{vd1}}{\beta_r^4} \qquad \text{(Eq 2-14)}$$

Where:

Variable	Definition or Description	Units US Customary (SI—Metric)
D_1	Reducer reduced pipeline (small-end) diameter	in. (mm)
D_2	Reducer large pipeline (large-end) diameter	in. (mm)
K_1	Reducer resistance coefficient based on the reduced (small-end) diameter, D_1	dimensionless
K_2	Reducer resistance coefficient based on the large-end diameter, D_2	dimensionless
K_{vd1}	Valve resistance coefficient based on the reference diameter, D_1 (typically the nominal diameter, D)	dimensionless
K_{vd2}	Valve resistance coefficient based on the diameter, D_2	dimensionless
L	Reducer end-to-end length	in. (mm)
α	Reducer (increaser) included angle, degrees; for angles ≤45 degrees ($\leq \pi/4$ rad)	degrees (rad)
β_r	Beta ratio for reducer flow resistance calculation: Reducer (increaser) diameter ratio	dimensionless

Reducer Assembly Flow Coefficient Example

Given the following input valve and reducer data, this example calculates the flow coefficients for the reducer, valve, and increaser as an assembly with the calculated data shown in Tables 2-1 and 2-2:

INHERENT AND INSTALLED CONTROL VALVE FLOW CHARACTERISTICS

From the definitions at the beginning of the chapter, the control valve flow characteristic is the percentage of maximum flow through the valve plotted against travel position. In the case of quarter-turn valves, the travel position may be either percentage of travel or opening angle. The inherent characteristic is that of the valve alone and is the percentage of full-open C_v plotted against travel position as shown in Figure 2-4. Once the valve is

Table 2-1 Reducer and increaser calculations

Input	Input	Input	Calc.	Calc.	Calc.	Calc.	Calc.	Calc.
Valve Size D, D_1 *in.* *(mm)*	Pipe Size D_2 *in.* *(mm)*	Fitting Length *in.* *(mm)*	Included Angle α rad Eq 2-9	Beta Ratio (D_1/D_2) β_r Eq 2-12	Reducer K_1 Valve Diameter Basis Eq 2-10	Increaser K_1 Valve Diameter Basis Eq 2-11	Reducer K_2 Pipe Diameter Basis Eq 2-13	Increaser K_2 Pipe Diameter Basis Eq 2-13
24 (600) (610)*	36 (900) (914)*	36 (900) (914)	0.330	0.667	0.07	0.13	0.37	0.67

* Nominal valve and pipe sizes may be based on 25 mm/in. rather than 25.4 mm/in.

Table 2-2 Valve, reducer, and increaser assembly adjusted flow coefficients

Input	Input	Input	Calc.	Calc.	Calc.	Calc.
Valve Angle θ *deg.*	$K_{v\theta}$ *	Valve $C_{v\theta}$ *gpm/psi½* *	Reducer Assembly Flow Resistance "$K_{ra1\theta}$" Valve Diameter Basis	Reducer Assembly Flow Resistance "$K_{ra2\theta}$" Pipe Diameter Basis	Reducer Assembly Flow Coefficient "$C_{vassy\theta}$" ("$C_{vmassy\theta}$") *gpm/psi½* (m^3/h) Eq 2-5 Eq 2-6	Inherent Characteristic, Percentage of Maximum Assembly Flow Coefficient "% C_{vmax}" ("% C_{vmmax}")
90	0.30	31,391	0.50	2.56	24,195 (20,808)	100.0
80	0.40	27,185	0.60	3.06	22,105 (19,010)	91.4
70	1.10	16,393	1.30	6.61	15,051 (12,944)	62.2
60	3.10	9,765	3.30	16.73	9,458 (8,134)	39.1
50	8.30	5,968	8.50	43.06	5,896 (5,071)	24.4
40	24.80	3,453	25.00	126.59	3,438 (2,957)	14.2
30	83.30	1,884	83.50	422.74	1,882 (1,619)	7.8
20	333.30	942	333.50	1,688.37	941 (809)	3.9
10	3,000.00	314	3,000.20	15,188.54	314 (270)	1.3

Calc.: calculation

* The valve data are for the example only and have no relationship to an actual valve.

installed into a system, the flow resistance of the system interacts with the valve, and the installed characteristic of the valve and system resistance changes as shown in Figure 2-5.

The graph lines of Figure 2-5 show how the valve and system interact as the system length and resistance increase. The system length is shown as the number of diameters of pipe (L/D). As shown in this figure, the installed characteristic curve moves up and to the left as the system resistance increases. If the valve is installed in a system by itself and there

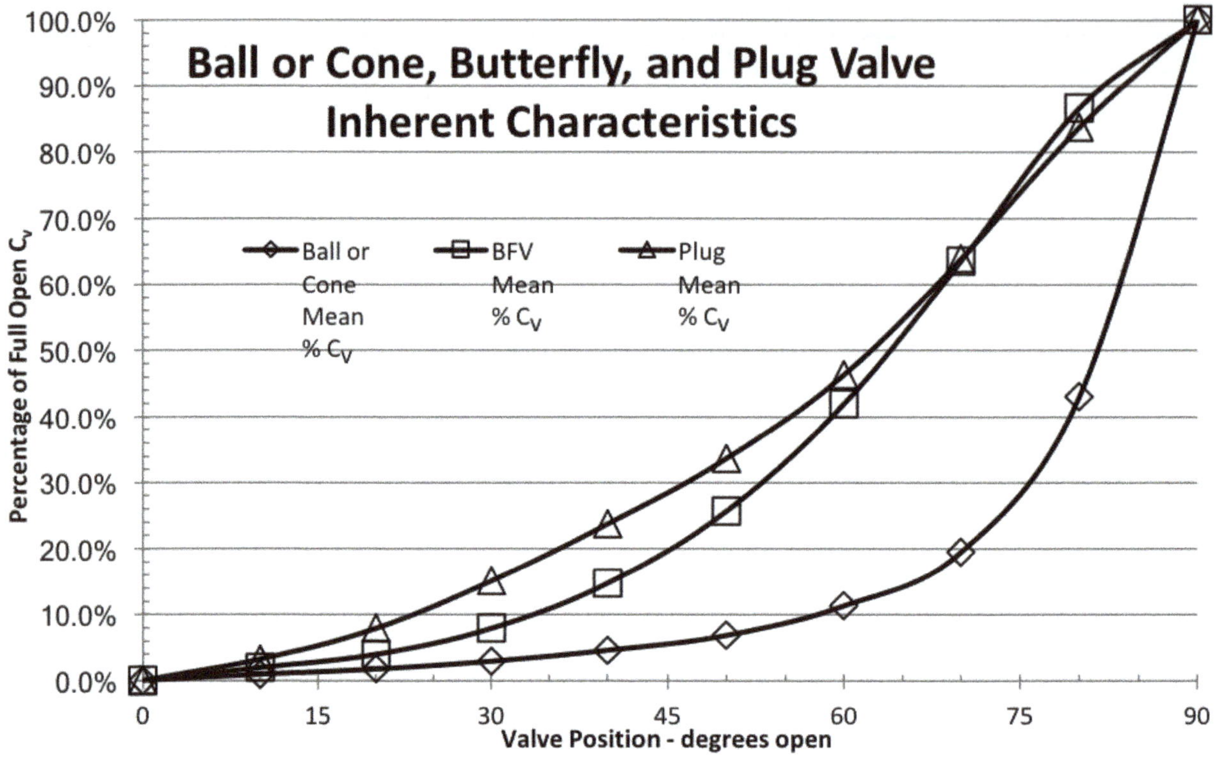

Figure 2-4 Typical inherent valve characteristics

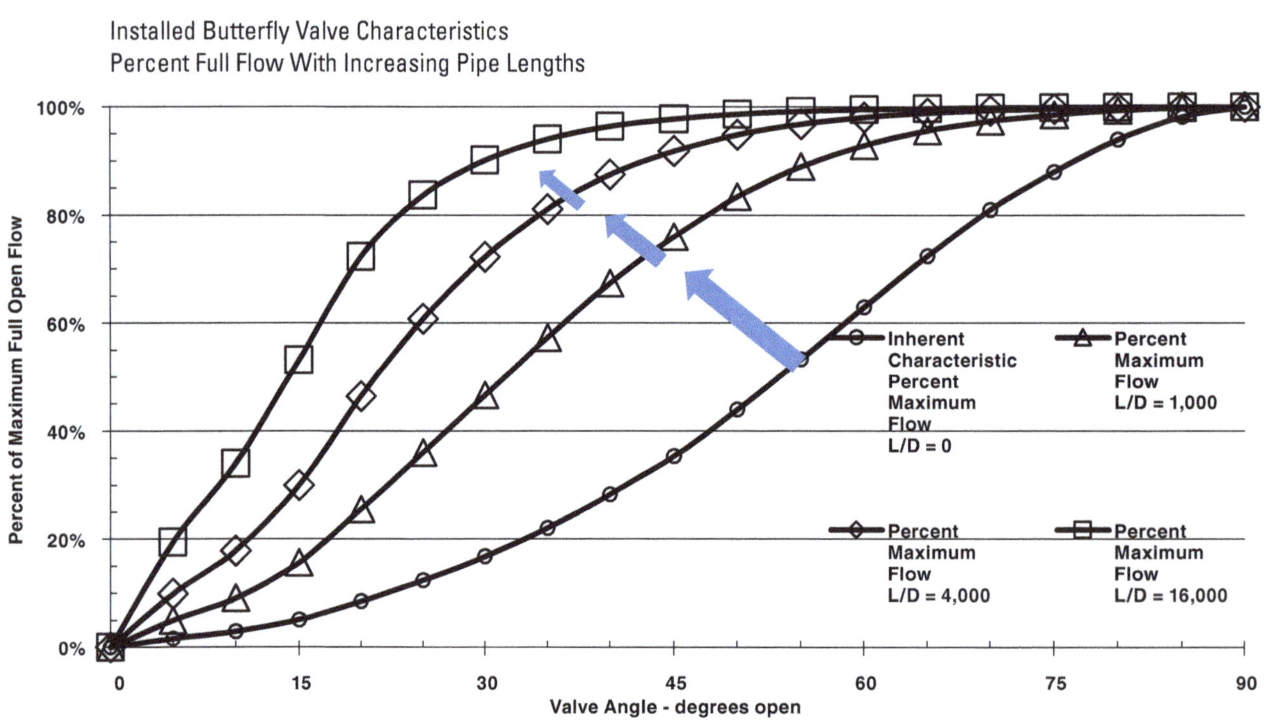

Figure 2-5 Installed valve characteristics

is little or no piping attached, the valve has complete control of the flow, and the differential pressure across the valve is essentially a constant at all angles. As the system resistance increases, the valve's capability to control the flow is reduced to lower valve angles.

FULL SYSTEM MODEL VERSUS EQUIVALENT RESISTANCE SYSTEM MODEL

To fully understand and predict system and valve performance during the stroking of the valve from open to close and the reverse would require a full and complete system and valve model. This can be arduous and requires full details of every valve, fitting, pipe, and fluid component in the system for every system lineup and flow path combination. The most accurate analysis is to build this complete and verified model for analyzing the valve's control capability and actuator sizing. During the design and procurement phases of a project, this is not practical or even possible.

The contributing members of AWWA C504 developed this equivalent resistance system model (ERSM) approach circa 1966 to afford a simple and practical method to design and predict pressure and flow conditions during valve operation as well as determine the valve's actuator size (strength and capability) requirements. In its simplest (constant-head) form, this model requires only two system inputs for a given valve. These are the maximum differential pressure (or head) when the valve is closed and the maximum flow rate (or line velocity) when the valve is fully open. From these data, calculating differential pressure, flow rate, and operating torque requirements at intermediate valve positions is possible. The basic assumption is that if the closed differential pressure (or head) and the full-open flow rate (or line velocity) are correct for the given system and lineup, the system resistance must be equal to Eq 2-15:

$$K_{sys} = \frac{2 \times g \times \Delta H_{max}}{V_{max}^2} - K_{v90}$$

The need for cavitation and choking predictions was later requested to assist the system designers with control purposes. Cavitation and choking are dependent on the fluid's absolute pressures upstream and downstream of the valve, the water temperature, and the valve's location within the system. This necessitates calculation of the line pressure just upstream of the valve as well as the differential pressure (or downstream line pressure) at each valve position. Therefore, the system input data for choking and cavitation evaluations require the closed line pressure just upstream of the valve, the differential pressure across the valve, the water temperature, and the location of the valve within the system. The valve's position within the system can be based on the estimated percentage of the system resistance that is upstream of the valve so that the system resistance can be estimated for each of the upstream and downstream system portions. If these additional data are not known (or at least estimated), the choking and cavitation conditions cannot be assessed. Performing multiple analyses may be necessary if there are multiple system lineups including various pressure sources, paths, and ending conditions.

This simplified ERSM is for use as a design tool and not a final analysis of as-installed conditions. For an accurate calculation of installed operating performance, one should use a more complete system model.

EQUIVALENT RESISTANCE SYSTEM MODEL

As can be seen from the previous installed characteristics discussions, the system resistance affects the valve's differential pressure and flow rate at a given valve position. To see how the valve controls the flow and pressure drop and to determine operating torque, the

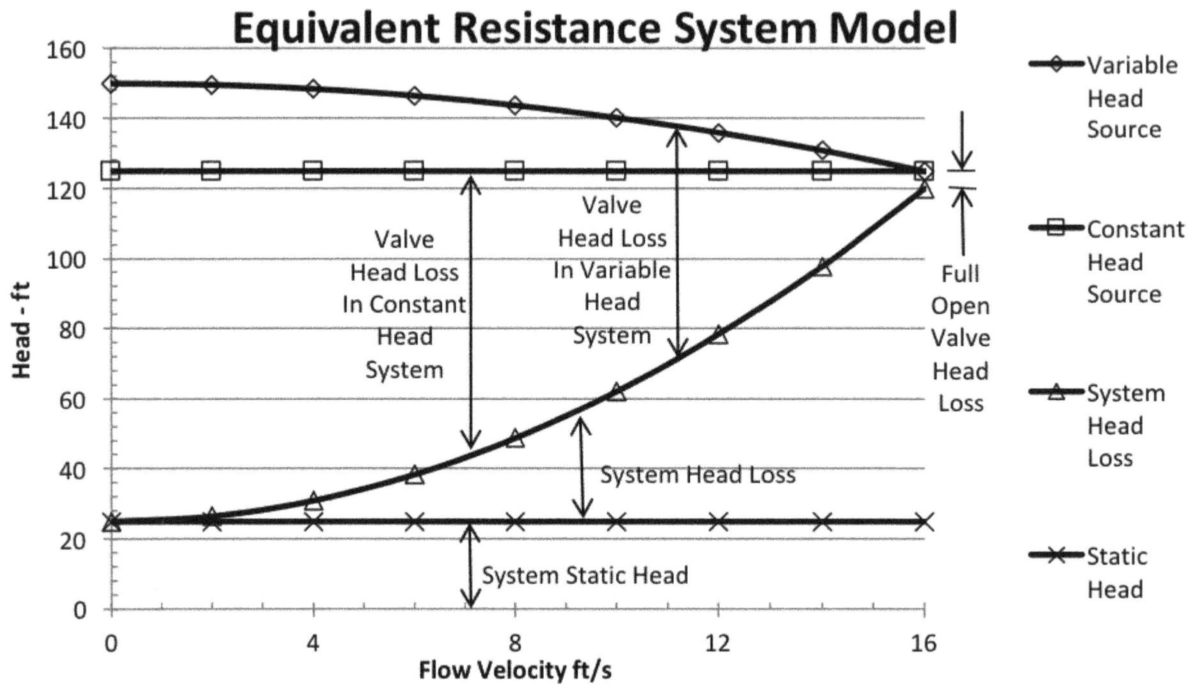

Figure 2-6 ERSM

installed characteristic of the system must be calculated. This simplified model is based on the maximum potential energy of the system (differential pressure) and the maximum kinetic energy of the system (flow). The maximum potential energy of the system is the differential head or pressure across the valve when it is closed (and the flow rate or velocity is zero). The maximum kinetic energy of the system is the maximum flow rate or line velocity when the valve is fully open. With these two inputs, an equivalent system resistance (K_{sys}) can be determined. This model is graphically shown in Figure 2-6.

This model assumes that during the valve's opening or closing operation, the remainder of the system is not changing and that the only flow resistance change is from the valve being analyzed. Most systems have several operating functions and conditions such as startup, shutdown, multiple normal operating conditions or scenarios, and emergency functions. Typically, the highest differential head or pressure and the greatest velocity or flow rate are used even if they do not coincide functionally. This is a conservative assumption for actuator sizing. Alternately, each operating case can be analyzed separately.

The valve's location within the system is not required for purposes of actuator sizing and classical flow valve head-loss calculations. As previously stated, for choking and cavitation analysis, determining the static pressure just upstream of the valve is necessary. For choking and cavitation analysis, the equivalent system resistance is split into upstream and downstream components, and the closed static upstream (or downstream) pressure is required as well as the differential pressure at the valve. It can be conservatively assumed that the valve is located at or near the end of the system, that most or all of the system losses occur upstream of the valve, and that the pressure downstream of the valve approaches 0 psig (0 kPa). This is a worst-case analysis, but results are often unfavorable or unacceptable, as choking and heavy cavitation are often predicted.

For single-point control valve sizing calculations, the upstream pressure, the controlled flow rate, the downstream pressure or differential pressure, and water temperature are required to predict the throttled valve's stroke position, choking, and the operating cavitation index. Given three or more sets of controlled flow conditions (two pressures, flow, and temperature) for the extremes of the process, "operating windows" can be determined, and performance can be anticipated within the boundaries of these windows. These windows are explained in Chapter 7.

Constant-Head Source Methodology

A typical constant-head source application in a water system is the flow of water from an elevated reservoir to a lower-elevation storage tank. Although the water level in the elevated reservoir may change throughout the day, at any given time the supply head (the elevation of the water level in the reservoir) is considered constant regardless of the water outflow rate. The sum of all other losses, except for the valve being analyzed, in a flow system equals ΔH_{sys}. Again, the calculation of operating torque for a single quarter-turn valve in a system assumes that no other variable loss coefficients, such as other valves, are changing in the system during the valve's travel. Therefore, the velocity head-loss coefficient of all components other than the valve in question can be considered constant and equal to K_{sys}. When calculating the head loss (or differential pressure), flow rate (or line velocity), and operating torque of a quarter-turn valve, the system will be evaluated as two basic components: (1) the valve being analyzed and (2) the remainder of the system piping. In a system with a constant-head source, ΔH_{sys} is considered constant for all flow conditions. When a valve in the system is closed, the maximum differential pressure across the valve (ΔH_{max}) is equal to the maximum differential across the system (ΔH_{sys}).

The system is modeled based on the maximum shutoff (closed) differential head or differential pressure and the maximum flow rate or velocity. (NOTE: These inputs are required by AWWA C504; AWWA C507; AWWA C516; AWWA C517; AWWA C519; and AWWA C522.)

The following steps and equations are used to develop the head or pressure losses and flow rates at various valve positions using the ERSM.

1. Calculate K_{sys} using K_v for a fully open quarter-turn valve:

$$K_{sys} = \frac{2 \times g \times \Delta H_{max}}{V_{max}^2} - K_{v90} \qquad \text{(Eq 2-15)}$$

Where:

Variable	Definition or Description	Units US Customary (SI—Metric)
g	Gravitational constant Acceleration due to gravity, 32.2 ft/s² (9.81 m/s²). NOTE: Three significant figures are practical for most calculations. Critical calculations and raw test data evaluations and reductions are best performed with greater accuracy.	ft/s^2 (m/s^2)
K_{sys}	System flow resistance coefficient (excluding the valve)	dimensionless
K_{v90}	Flow resistance coefficient of valve at full open (≈90 degrees, $\approx \pi/2$ rad). NOTE: Use of K_{V90} assumes the valve travels 90 degrees to full open.	dimensionless
V_{max}	Maximum full-open velocity ft/s(NOTE: based on nominal valve diameter if converted from a quantity flow rate)	feet per second, ft/s, or ft/s (meters per second, mps, or m/s)

Variable	Definition or Description	Units US Customary (SI—Metric)
ΔH_{max}	Head loss across the closed valve or total system with valve closed	feet of water (meters of water)
ΔH_{sys}	Head loss across the system	feet of water (meters of water)

2. The flow velocity through the valve at the valve angle (θ) may be calculated using:

$$V_\theta = \sqrt{\frac{2 \times g \times \Delta H_{max}}{(K_{sys} + K_{v\theta})}} \quad \text{(Eq 2-16)}$$

Where:

Variable	Definition or Description	Units US Customary (SI—Metric)
$K_{v\theta}$	Flow resistance coefficient of the valve at valve angle θ	dimensionless
V_θ	Approach velocity of fluid flow at valve angle θ	feet per second, ft/s, or ft/s (meters per second, mps, or m/s)

3. Calculate $\Delta H_{v\theta}$ at the valve angle (θ):

$$\Delta H_{v\theta} = \frac{K_{v\theta} \times V_\theta^2}{2 \times g} \quad \text{(Eq 2-17)}$$

or:

$$\Delta H_{v\theta} = \frac{\Delta H_{max} \times K_{v\theta}}{(K_{v\theta} + K_{sys})} \quad \text{(Eq 2-18)}$$

Where:

Variable	Definition or Description	Units US Customary (SI—Metric)
$\Delta H_{v\theta}$	Head loss across the valve at angle θ	feet of water (meters of water)

4. Calculate ΔP_θ at the valve angle (θ):

$$\Delta P_{v\theta} = 0.4335 \times \Delta H_{v\theta} \quad \text{(Eq 2-19)}$$

or:

$$\Delta P_{v\theta} = \frac{\Delta P_{max} \times K_{v\theta}}{(K_{v\theta} + K_{sys})} \quad \text{(Eq 2-20)}$$

Where:

Variable	Definition or Description	Units US Customary (SI—Metric)
$\Delta P_{v\theta}$	Pressure drop (or loss) across the valve at valve angle θ	psid (kPa)

5. Repeat steps 2–4 for other valve angles (θ).

Constant-Head Source Methodology Example

Given the following valve and system data:

- A 24-in. (600-mm) AWWA Class 150B BFV with 24-in. (600-mm) diameter single-offset disc and seat-side flow is used.
- Maximum head differential (ΔH_{max}) is 100 ft of water (ΔP_{max} = 43.4 psid [299 kPa-d], Eq 2-19).
- Closed upstream head ($H_{upstream}$) is 200 ft of water ($P_{upstream}$ = 86.7 psi [598 kPa], Eq 2-19).
- System downstream static head ($H_{dnstream}$) is $H_{upstream} - \Delta H_{max}$ = 200 – 100 = 100 ft of water ($P_{dnstream}$ = 43.4 psi [299 kPa], Eq 2-19).
- Maximum system flow rate, Q_{max}, = 20,000 gpm (V_{max} = 14.2 ft/s [4.33 m/s], Eq 2-7).
- Water temperature is 60°F (15.6°C) (with a vapor pressure of 0.58 ft [0.177 m] absolute = 0.25 psia [1.72 kPa] = –33.3 ft [–101 m] gauge, –14.4 psig [–99.3 kPa]).
- $K_{v\theta}$ is listed in Table 2-3.
- $K_{sys} = \dfrac{2 \times 32.17 \times 100}{14.2^2} - 0.3 = 31.61$ from Eq 2-15.
- Percentage of system upstream of valve = 75 percent ($K_{sys\text{-}up}$ = 23.70).
- See Figure 2-7 for system model as well as Table 2-3 and Figure 2-8 for calculation results and a graph of the results.

Table 2-3 Head-loss calculation data for constant-head source example

Input	Input	Calc.	Calc.	Calc.	Calc.	Calc.	Calc.	Calc.
Valve Angle θ *deg.*	$K_{v\theta}$ *	$\Delta H_{v\theta}$ *ft* *(m)* †	V_θ *ft/s* *(mps)* †	ΔH_{sys} (100%) *ft* *(m)* ‡	ΔH_{SYS} Upstream of Valve (75%) *ft* *(m)* §, ¶	Velocity Head at Valve $V^2/2G$ *ft* *(m)* §, ¶	Static Head H Upstream of Valve *ft* *(m)* §, ¶	Static Pressure *P* Upstream of Valve psi *(kPa)* §, ¶
90	0.30	0.9 (0.27)	14.2 (4.33)	99.1 (30.2)	74.3 (22.6)	3.1 (0.95)	122.6 (37.4)	53.1 (366)
80	0.40	1.2 (0.37)	14.2 (4.33)	98.8 (30.1)	74.1 (22.6)	3.1 (0.95)	122.8 (37.4)	53.2 (367)
70	1.10	3.4 (1.04)	14.0 (4.27)	96.6 (29.4)	72.5 (22.1)	3.1 (0.95)	124.5 (37.9)	54.0 (372)
60	3.10	8.9 (2.71)	13.6 (4.15)	91.1 (27.8)	68.3 (20.8)	2.9 (0.88)	128.8 (39.3)	55.8 (384.7)

(continued)

Table 2-3 Head-loss calculation data for constant-head source example *(Continued)*

Input	Input	Calc.	Calc.	Calc.	Calc.	Calc.	Calc.	Calc.
Valve Angle θ *deg.*	$K_{v\theta}$ *	$\Delta H_{v\theta}$ *ft* *(m)* †	V_θ *ft/s* *(mps)* †	ΔH_{sys} (100%) *ft* *(m)* ‡	ΔH_{SYS} Upstream of Valve (75%) *ft* *(m)* §, ¶	Velocity Head at Valve $V^2/2G$ *ft* *(m)* §, ¶	Static Head H Upstream of Valve *ft* *(m)* §, ¶	Static Pressure *P* Upstream of Valve psi *(kPa)* §, ¶
50	8.30	20.8 (6.34)	12.7 (3.87)	79.2 (24.1)	59.4 (18.1)	2.5 (0.76)	138.1 (42.1)	59.9 (413)
40	24.80	44.0 (13.4)	10.7 (3.26)	56.0 (17.1)	42.0 (12.8)	1.8 (0.55)	156.2 (47.6)	67.7 (466)
30	83.30	72.5 (22.1)	7.5 (2.29)	27.5 (8.38)	20.6 (6.28)	0.9 (0.27)	178.5 (54.4)	77.4 (534)
20	333.30	91.3 (27.8)	4.2 (1.28)	8.7 (2.6)	6.5 (2.0)	0.3 (0.09)	193.2 (58.9)	83.8 (578)
10	3,000.00	99.0 (30.2)	1.5 (0.46)	1.0 (0.31)	0.8 (0.24)	0.0 (0.0)	199.2 (60.7)	86.3 (595)
0	Inf.	100.0 (30.5)	0.0 (0.0)	0.0 (0.0)	0.0 (0.0)	0.0 (0.0)	200.0 (61.0)	86.7 (598)

Calc.: calculation

* Valve data are generic for the example only and have no relationship to an actual valve.
† $\Delta H_{v\theta}$ is used for head-loss (pressure loss), velocity (flow), and torque calculations.
‡ ΔH_{sys} is informational only.
§ These calculations are used for choking and cavitation calculations and not required for head loss (pressure loss), velocity (flow), and torque calculations.
¶ These columns are calculated using standard hydraulic principles left to the reader. (See Crane [2018] and others.)

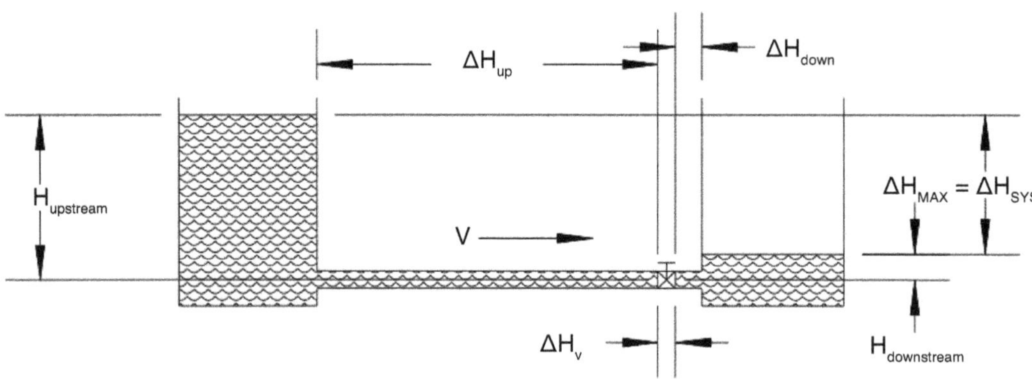

Figure 2-7 Example ERSM diagram

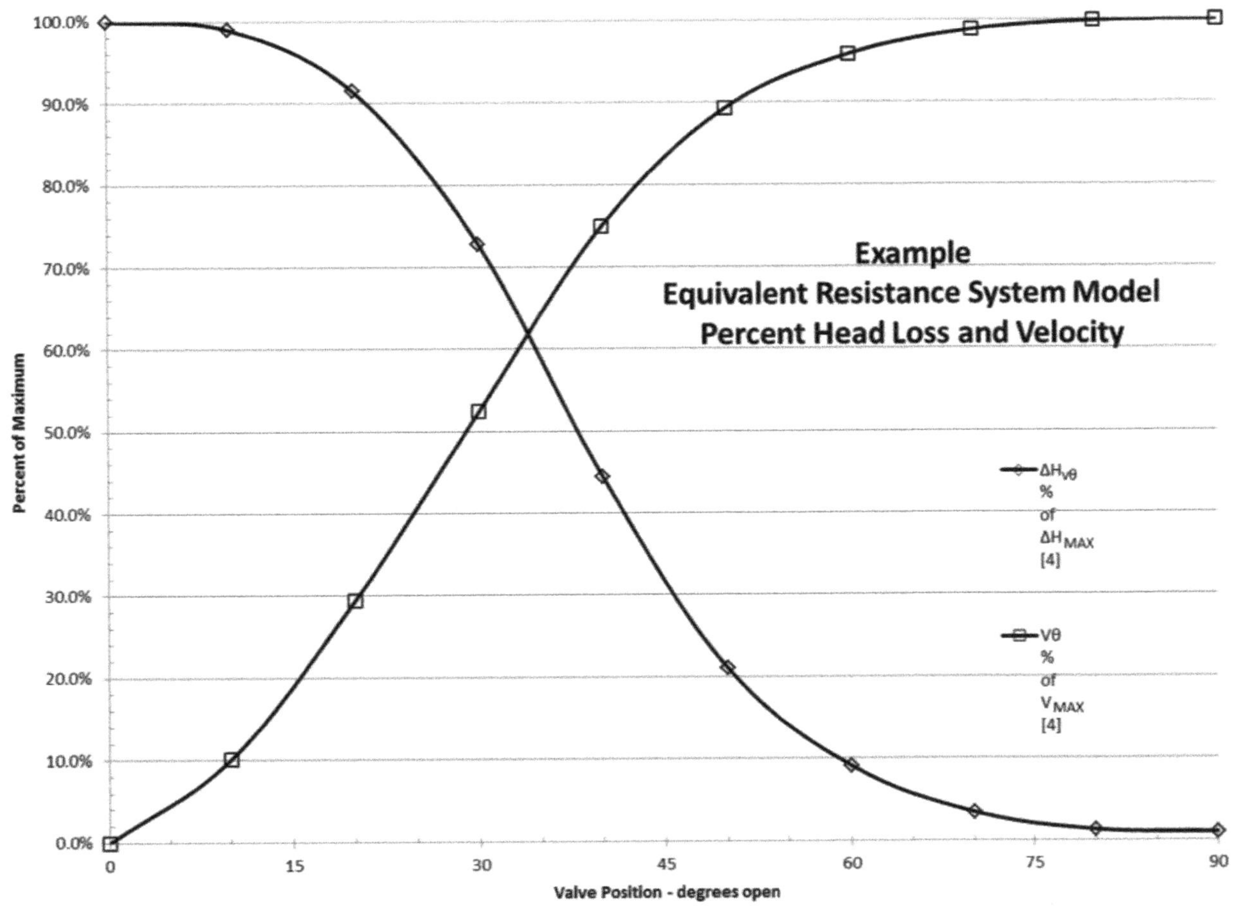

Figure 2-8 Graphical relationship between head loss and velocity

CHOKING AND CAVITATION INPUTS AND CALCULATION

Many water works valve analyses may be performed without regard to choking and cavitation limitations. To examine for choking and cavitation at the valve (midstroke positions), dividing the system resistance (K_{sys}) into upstream (K_{sysup}) and downstream (K_{sysdn}) components is necessary. This division permits the calculation of the pressure just upstream of the valve, the differential pressure across the valve, the fluid velocity or flow rate, and the various operating torque components at each valve position.

The inputs of closed static upstream pressure, water temperature, and percentage of the system upstream of the valve are also needed in addition to the closed-valve differential pressure and maximum full-open flow rate (or velocity). This simplified methodology uses a percentage of the system that is upstream of the valve ($\%K_{sysup}$) as an input into the angle-by-angle midstroke calculations. This is a practical engineering estimate without having full system details at hand. Once implemented, to evaluate the effect and sensitivity of this assumption by performing several calculations changing different upstream percentages is reasonably simple. The system should be completely hydraulically modeled in its entirety for highest accuracy.

The valve's cavitation and choking coefficient data are also necessary. These calculations are based on the pressure and flow input data only. Several different system lineup conditions may need to be calculated for better understanding of the range of operating conditions.

The choked differential at each valve position is calculated using the following:

$$\Delta P_{choke\theta} = F_{L\theta}^2 \times \left(P_{u\alpha} - F_f \times P_{v\alpha}\right) \qquad \text{(Eq 2-21)}$$

Where:

Variable	Definition or Description	Units US Customary (SI—Metric)
F_f	Liquid critical pressure ratio factor (NOTE: For water in the range of 32°F [0°C] to 125°F [51.7°C], water $F_f \approx 0.96$.)	dimensionless
$F_{l\theta}$, $F_{l\theta}^2$	Liquid pressure recovery factor of a valve without attached fittings. This experimentally determined factor depends on the type of valve and its internal geometry. Figure 2-9 shows a graphical representation of typical values.	dimensionless
P_{ua}	Reference pressure just upstream of the valve for choking and cavitation analysis*	psia (kPa)
P_{va}	Absolute vapor pressure (examples: P_{va} = 0.0867 psia [5.97 kPa] for water at 32°F [0°C] and P_{va} = 1.93 psia [13.3 kPa] for water at 125°F [51.7°C] measured at sea level)*	psia (kPa)
$\Delta P_{choke\theta}$	Differential pressure at onset of choked flow†	psid (kPa)

* Pressure may be in any consistent units.

† As F_f is very close to 1.0 for water within the range of temperatures between 32°F (0°C) and 125°F (51.7°C), the choking formula may be reduced to: $\Delta P_{choke} = F_l^2 \times (P_{u\alpha} - P_{v\alpha})$.

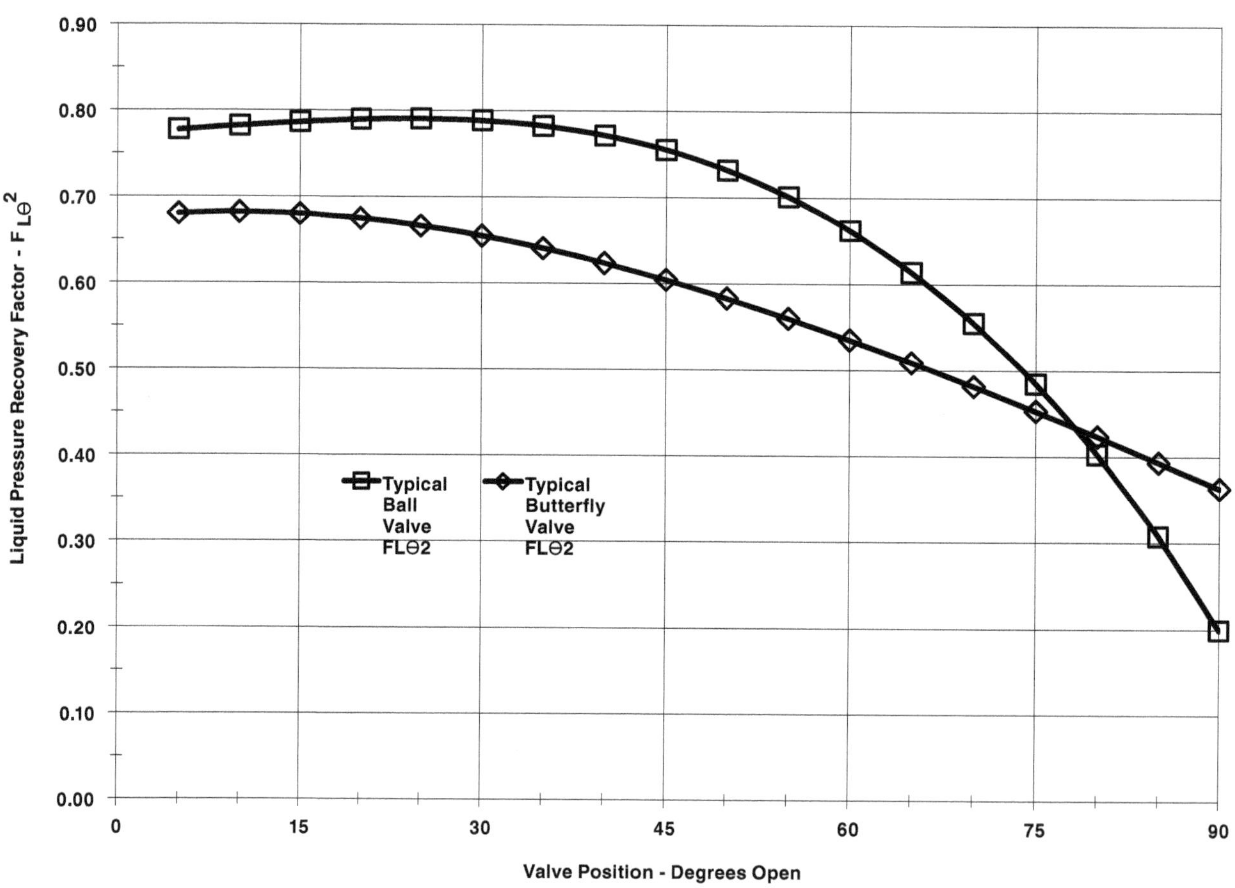

Figure 2-9 Graphical relationship of typical BFV and BV $F_{l\theta}^2$ values

VARIABLE HEAD SOURCE METHODOLOGY

Similar to the constant-head source methodology, a typical variable head source application is pumping water from a source (such as a tank, reservoir, or lake) to an output elevation or pressure to a vessel or reservoir. In this application, the supply head of the pump is a function of the flow rate through the pump, as provided on the pump's head curve. The sum of the head losses in a flow system at any given time equals ΔH_{sys}. This method is seldom necessary, as assuming a constant-head source is conservative for actuator sizing use. If the pump curve is a second-order inverted parabola (similar to that shown in Figure 2-6), the results from a constant-head assumption and the variable head methodologies are identical. This variable head method is only necessary when the pump curve is oddly shaped and cannot be easily approximated by a second-order polynomial. This is because the variable shutoff head (or pressure) is higher than the constant-head shutoff head (or pressure) for the same run-out or maximum fluid flow (or velocity). Therefore, using the constant-head method, the calculated system resistance (K_{sys}) will increase above the system resistance calculated for the variable head curve. (NOTE: This increase is essentially equal and opposite to the reduction in the second-order source head curve.) As the valve responds to the area between the head (or pressure) line or curve and the system resistance curve, the resulting calculations match. The fluid flow rate or velocity and the valve's head loss (difference between the system curve and the variable head source curve) define the valve's position needed. Basically, the method is to find the necessary valve resistance at a given flow and head-loss pair on the pump curve, which defines the valve's position (or angle). Alternately, curve-fit functions of the valve's head-loss and torque coefficients and position can be developed for interpolation to intermediate positions.

1. Calculate K_{sys} using Eq 2-15 with one exception as noted here.

The ΔH_{sys} at V_{max} are read from the pump curve at the maximum full-open flow rate or velocity. The value for ΔH_{sys} is the difference between the pump supply head and the system static head.

2. Calculate the flow velocity of the system at the desired valve angle (θ) using:

$$V_{\theta} = \sqrt{\frac{2 \times g \times \Delta H_{max}}{\left(K_{sys} + K_{v\theta}\right)}} \qquad \text{(Eq 2-22)}$$

In this equation, $K_{v\theta}$ is the K of the valve at the desired valve angle. Since $\Delta H_{sys\theta}$ is variable and dependent on $V_{V\theta}$, this equation must be solved using an iterative process.

a. Assume a value for $V_{v\theta}$ less than that of V_{θ} at the next higher valve angle.
b. Calculate the corresponding $\Delta H_{sys\theta}$:

$$\Delta H_{sys\,\theta} = \frac{\left(K_{sys} + K_{v\theta}\right) \times V_{\theta}^{2}}{2 \times g} \qquad \text{(Eq 2-23)}$$

c. Examine the pump curve and determine whether $\Delta H_{sys\theta}$ matches the curve.
d. If the $\Delta H_{sys\theta}$ is not a good match with the calculated value, then assume a higher or lower V_{θ} and repeat steps 2a–2c until $\Delta H_{sys\theta}$ matches the pump and system curve.

3. Calculate $\Delta H_{v\theta}$ at the desired valve angle (θ):

$$\Delta H_{v\theta} = \frac{K_{v\theta} \times V_{\theta}^{2}}{2 \times g} \qquad \text{(Eq 2-24)}$$

4. Calculate $\Delta P_{v\theta}$ at the desired valve angle (θ):

$$\Delta P_{v\theta} = 0.4335 \times \Delta H_{v\theta} \quad \text{(Eq 2-25)}$$

5. Repeat steps 2–4 for other valve angles.

COMPARISON OF CONSTANT AND VARIABLE HEAD RESULTS

The comparison for a constant-head source and a second-order pump curve fit similar to that in Figure 2-6 with a single curvature orientation shows that the calculated valve head losses are identical at all valve positions and fluid velocities (or flow rates).

The comparison for a constant-head source and third-order (or higher) pump curve fit (variable positive and negative head graphical curvature) shows that the valve head (pressure) losses are practically identical at all positions where the pump head reduces at a regular rate.

At the valve positions where the head loss is not consistently trending downward with increased flow or velocity (in a manner similar to that of Figure 2-6) the valve's head loss at the indicated flow rate is less than that calculated from the constant-head model. Therefore, the constant-head model is conservative for actuator sizing determinations but has an error in control valve positioning evaluations, cavitation, and choking estimates.

The constant-head source method is more direct and less rigorous but may contain errors in control valve calculations at lower valve positions. Control valve sizing, operating window calculations, and positioning sensitivity evaluations are developed in Chapter 7.

The methods necessary to understand valve performance as a control element are not appropriate for actuator sizing. The specific system conditions for control sizing are not necessarily the worst-case, highest-energy conditions of operation that should be used for actuator sizing evaluations.

The use of the pump curves and the velocity calculation methodology does provide greater accuracy when pumps are involved but requires additional inputs and alternate enhanced calculation procedures. This procedure represents operation at only one system condition and may need multiple evaluations. Pump curve data and full system data are not always available during the design phases of a project, so the constant-head model is appropriate.

Actuator sizing evaluations should be performed at the maximum possible valve differential head (pressure) and maximum possible full-open fluid velocity (or flow rate) regardless of whether a pump curve is available. Systems are seldom functioning at their combined extreme limits. The midstroke performance flow and differential head (pressure) results and attributes are not applicable to other lesser system operating and boundary conditions.

CONTROL VALVE APPLICATIONS

Control valve applications need to be analyzed over the range of maximum, normal, and minimum operating (boundary) head (pressure) and fluid velocity (flow rate) conditions, resulting in a matrix of several operating system lineups and requirements.

When pumps are involved, evaluation of the throttling or modulating conditions using pump curve data and at several individual (high and low) system boundary conditions, including high and low system resistances, is more accurate. These analyses should be performed at the extremes of anticipated operating boundary conditions and demand requirements. This is performed not as a full stroke (of unnecessary torque and actuator sizing data) but on a matrix of the individual system operating, system resistances, and performance requirements. These operational matrixes provide an operating window of system and valve control ability. The control valve sizing, positioning, and sensitivity evaluations should be based on a series of system boundary conditions (not a single worst-case actuator sizing review). Valve performance at positions other than those required to maintain the flow or pressure control conditions and evaluate choking and cavitation are seldom necessary.

When each flow condition in the operational matrix is presented graphically in differential pressure, flow, and cavitation windows, the extent of the valve's control capability can be visually evaluated. Control valve selection, sizing, and window development are addressed in Chapter 7.

ENERGY CALCULATIONS

One reason for using a quarter-turn valve is that they offer low head loss in the full-open position than many control or globe valves and offer a compact size. In a pumped system, when flow passes through a valve or fitting, the resulting head loss requires additional energy from the pumps. Valve head loss therefore translates directly into higher energy consumption by the pump motors.

An equation used for evaluating yearly energy cost is as follows:

(In US customary units);
$$A_{cost} = \frac{1.65 \times Q \times \Delta H \times S_g \times C \times U}{E} \quad \text{(Eq 2-26)}$$

(In SI metric units);
$$A_{cost} = \frac{23.8 \times Q \times \Delta H \times S_g \times C \times U}{E} \quad \text{(Eq 2-27)}$$

Where:

Variable	Definition or Description	Units US Customary (SI—Metric)
A_{cost}	Annual energy cost	$/year
C	Cost of electricity	$/kWh
E	Efficiency of pump and motor set (80 percent; 0.80, typical)	decimal/fraction (%/100)
Q	Volumetric flow rate	gpm (m^3/h)
S_g	Specific gravity of liquid relative to water at 60°F (16°C) (water = 1.0)	dimensionless
U	Pump usage percentage, 100 percent (1.0) equals 24 h per day	decimal/fraction (%/100)
ΔH	Head loss between any two reference points in a system	feet of water (meters of water)

Energy Calculation—Example

For example, in a 24-in. (600-mm) BV or RCV (K_v = 0.05±0.01 [±20 percent]), a BFV or PV (K_v = 0.50±0.1 [±20 percent]), and a globe valve (K_v = 4.00±0.8 [±20 percent]) installed in a 15,000-gpm (10.6-ft/s, 3.2-m/s, 1-m^3/s) pump line, the efficiency of the pump and motor set is 80 percent (0.8, decimal/fraction), and the cost of electricity is $0.09 per kWh. The cost of the valves head loss is computed using Eq 2-26. The calculated data are shown in

Table 2-4 Pumping cost example calculation

Valve Type	Full-Open K_v	Full-Open ΔH at 15,000 gpm (3,407 m³/h) 10.6 ft/s (3.23 mps)	A_{cost} $
Ball or rotary cone 1	0.04	0.07	97.25
Ball or rotary cone 2	0.05	0.09	121.56
Ball or rotary cone 3	0.06	0.10	145.87
Butterfly or plug 1	0.40	0.70	972.50
Butterfly or plug 2	0.50	0.87	1,215.62
Butterfly or plug 3	0.60	1.05	1,458.75
Globe 1	3.20	5.59	7,779.98
Globe 2	4.00	6.99	9,724.97
Globe 3	4.80	8.38	11,669.97

Table 2-4. A 24-in. (600-mm) valve in the example can consume from $97 to almost $12,000 in annual electricity, assuming water ($S_g = 1.0$) is pumped through the valve for 50 percent (0.5, decimal/fraction) of the time.

This example is theoretically correct given the usage, flow rate, and the pump efficiency are assumed constant. In operation, the lower-loss valves allow greater pumped flow at the same system operating system pressure; the pump runs further out on the pump curve and may even require the pump to draw a different (slightly lower or greater) current during operation. The savings are truly seen in the usage (running time reduction) and increased pump efficiency to deliver the same quantity of water (or inventory). The savings will be seen in the pump usage reduction over time and reduced number of pumps operating at the same time to produce the system delivery demand.

As presented in Figure 2-10, the greatest energy cost change is between valve types and not between the relative resistances of similar valve models, classes, and types. This is not to say that globe valves in the example should not be used in pump-control

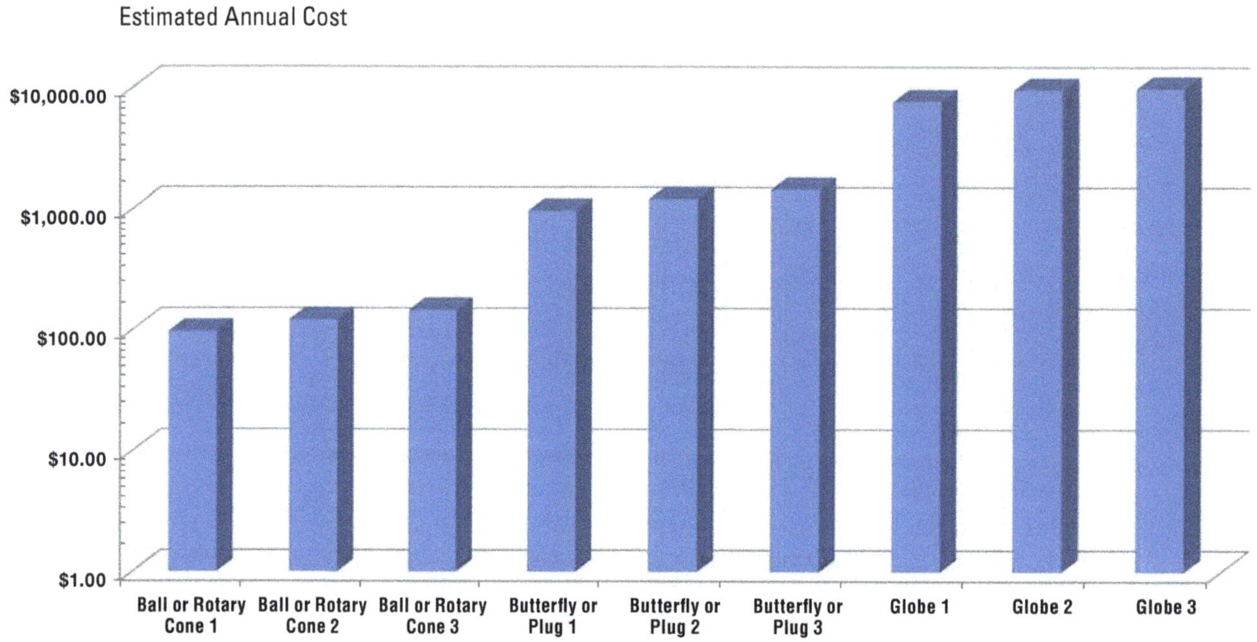

Figure 2-10 Comparative operating costs by valve type

service, as they offer different engineering performance variables, including a wider flow-control range, cavitation performance and trim variations, different inherent flow characterizations, and transient control solutions, but to point out that there is an energy cost associated with those performance differences. When looking at the same valve type, the savings between individual product models, pressure classes, or suppliers are generally insignificant and seldom measurable in service. For these low-loss quarter-turn valves, the difference between various suppliers is generally less than that of a single elbow or even the difference between short- and long-radius elbow.

REFERENCES

ANSI/AWWA. 2023. C504. Rubber-Seated Butterfly Valves. Denver: AWWA.

ANSI/AWWA. 2023. C507. Ball Valves, 4 in. Through 60 in. (150 mm through 1,500 mm). Denver: AWWA.

ANSI/AWWA. 2022. C522. Rotary Cone Valves, 6 In. Through 60 In. (150 mm through 1,500 mm). Denver: AWWA.

ANSI (American National Standards Institute)/AWWA. 2021. C516. Large Diameter Rubber-Seated Butterfly Valves, 78 in. (2,000 mm) and Larger. Denver: AWWA.

ANSI/AWWA. 2021. C516. Large-Diameter Rubber-Seated Butterfly Valves, Sizes 78 In. (2,000 mm) and larger. Denver: AWWA.

ANSI/AWWA. 2021. C517. Resilient-Seated Cast-Iron Eccentric Plug Valves. Denver: AWWA.

ANSI/AWWA. 2018. C519. High-Performance Waterworks Butterfly Valves—3 In. (75 mm) through 60 In. (1,500 mm). Denver: AWWA.

American National Standards Institute (ANSI)/International Society of Automation (ISA) S75.01.01. Flow Equations for Sizing Control Valves. Research Triangle Park, N.C.: ISA.

ANSI/ISA (International Society of Automation). 2008. S75.02.01-2008. Control Valve Capacity Test Procedures. Research Triangle Park, N.C.: ISA.

Crane. 2018. "Flow of Fluids Through Valves, Fittings and Pipe, Technical Paper No. 410." Crane.

This page intentionally blank.

Chapter 3

Valve Torque

In a quarter-turn valve, valve torque is the turning effort needed to rotate the valve's closure member (ball, cone, disc, or plug) or hold it in position. Torque varies with system conditions, valve design, and closure member position. The methodology given in this manual is a step-by-step procedure for predicting valve operating torque and represents the current method used by many quarter-turn valve manufacturers in the water industry.

Rotary cone valves (RCVs) are unique in their seating and unseating as they change from torque and rotation to lift, thrust, and axial movement. This involves the development of a series of thrust and screw torque to thrust conversion equations that are not currently included in this manual. Therefore, seating and unseating of the RCV are not covered until the linear valve manual of standard practice is available.

DISCUSSION OF TORQUE CALCULATIONS

The torque calculations are broken into 10 separate torque components, and each is derived from a first-principles approach. The 10 separate torque components are classified into two categories: (1) passive or friction-based or (2) active or dynamically generated. These 10 components are listed in Table 3-1.

Each of these separate torque components is evaluated mathematically from a first-principles approach, and their equations are presented, except for the buoyancy torque (item 7) and the thrust bearing torque (item 5). These two are generally considered as negligible for this scope of the valves. The components of hub seal friction torque (item 3), weight and center-of-gravity torque (item 6), lateral offset or eccentricity torque (item 8), and hydrostatic-unbalance torque (item 10) may not be applicable depending on the valve design and installation variables. Also, packing friction and hub seal friction may be considered as a single torque. Seating (and/or unseating) friction torque (item 1), packing friction torque (item 2), bearing friction torque (item 4), and dynamic or fluid dynamic torque (item 9) should always be included in operating torque calculations for ball valves

Table 3-1 Torque component category

Item No.	Torque Component	Torque Category
1	Seating (and/or unseating) friction torque	
2	Packing friction torque	
3	Hub seal friction torque	Passive or friction-based
4	Bearing friction torque	
5	Thrust bearing friction torque	
6	Weight and center-of-gravity torque	Active or dynamically generated
7	Buoyancy torque	
8	Lateral offset or eccentricity torque	
9	Dynamic or fluid dynamic torque	
10	Hydrostatic unbalance torque	

(BVs), butterfly valves (BFVs), and plug valves (PVs). The lift, thrust, and torque to thrust conversions for the seating and unseating of RCV are not evaluated, but all other midstroke torque requirements should be evaluated.

The passive torque components are friction related and in general are either constant for a given valve or directly dependent on the differential pressure. These components always oppose actuator motion and are generally considered to be essentially the same magnitude in either direction of operation (opening or closing), except for seating and unseating. Seating and unseating torque may be evaluated separately or considered the same when differences are small.

The active or dynamic torque components are generated in the valve by the effects of the internal fluid media (water) or gravity acting on the valve. These components may oppose or assist the actuator's operation. Because dynamic torque generally tends to close the valve, the actuator may act as a brake to control the speed of the closing stroke but must also overcome this torque in the opening stroke.

The separate actuating torque methodology provided here is generally used for valves of larger sizes. Actuator sizing in valves 12 in. (300 mm) and smaller is driven primarily by the passive/friction-based torque requirements, as the active torque components are a small fraction of the total required operating torque. The transition point at which the dynamically generated torque components become a major part of the total required torque depends on many factors of the valve design. However, it can be generally stated that this transition occurs in the 14-in. (350-mm) to 30-in. (900-mm) range for this scope. Actuator sizing for valves larger than 30 in. (900 mm) is often significantly based on the dynamic flow conditions.

Because of this and because the smaller-sized valves are easily tested and grouped into a smaller range of required actuator torque over the full span of the design pressure and flow rate, this complex calculation methodology may be replaced by a simple calculation based on size and pressure using curve-fitting techniques of test data. In the smaller sizes, the manufacturer may provide curve-fit equations or graphical or tabulated information.

This separate-effects methodology becomes increasingly important in the larger valve sizes, such as ≈18 in. (450 mm) and larger, and at very high fluid line velocities (greater than 16 ft/s or 4.9 m/s). To test many large-diameter valves is economically or physically infeasible; and using separate-effects calculations, model test data, and grouping of the test data is necessary. The modeling techniques of dimensionless coefficients, hydraulic similitude, grouping, interpolation, and extrapolation are not discussed in this manual.

The friction-based torque components are either constant or related to the valve diameter to the second power (D^2). Because the dynamic torque component is a function of the valve diameter to the third power (D^3), it becomes the major torque affecting the actuator sizing of the larger-size valves. This is why the maximum operating flow rate or fluid line velocity is needed for actuator sizing of larger valve sizes.

The hydrostatic-unbalance torque is also of great importance (if it exists) to larger valve sizes because it is a function of the diameter to the fourth power (D^4), although it can be ignored as insignificant in valves ≈36 in. (900 mm) and smaller. It is seldom seen under actual operating conditions, but its influence can be very significant in valve sizes larger than 36 in. (900 mm) when it is present.

This separate-effects methodology is then best applicable for determining the required actuator torque for the larger sizes of valves where the torque components are determined individually (rather than by curve-fitting techniques of the total torque) as the combination of both the operating shutoff differential pressure and maximum operating flow rate (or line velocity) has a significant effect on results.

The methods of calculating the required actuator torque, system flow, pressure drop, and cavitation indexes described here are applicable to most quarter-turn valves. The test-developed coefficients must be specific for the valve type and design, but the first-principles methodology is basically unchanged.

The flow and torque coefficient data (K_v or C_v and C_t) used must be from the same valve design and same test procedure. Care should be taken not to use the flow coefficients from one valve design and apply them to another design or mix flow and torque coefficients. In other words, the flow and torque coefficients should be a matching set of data.

DEFINITIONS

The methodology that follows is based on several terms and concepts that are defined in this section.

Torque Coefficients, C_t and $C_{t\theta}$

Torque coefficients are developed on the installation of a valve in a straight run of pipe without upstream or downstream flow disturbances, such as nearby elbows, tees, or increasers. The effects of these pipe fittings are beyond the scope of this manual, and such conditions should be brought to the attention of the valve manufacturer. Specific installation guidelines are given in Chapter 6. Manufacturers may consider torque coefficients to be proprietary information.

Bearing Torque, T_b, and $T_{b\theta}$

Bearing torque calculations are dependent on the valve closure member and shaft diameters and the coefficient of friction between the shaft or trunnion and the bearing materials. Minimum shaft diameters are listed in American National Standards Institute (ANSI)/AWWA C504, Rubber-Seated Butterfly Valves. A methodology for sizing minimum shaft diameter is given for larger valves in ANSI/AWWA C516, Large Diameter Rubber-Seated Butterfly Valves, 78 in. (2,000 mm) and Larger. Consult the valve manufacturer's drawings and other documentation for actual diameters or take measurements.

Bearing Coefficient of Friction, C_f

The shaft bearing material supports the shaft and disc in the valve body, allowing rotation. The static coefficient of friction for the bearing and shaft or trunnion material couple is needed to calculate valve bearing friction torque. Consult the valve or bearing manufacturer or other mechanical engineering references and handbooks for typical friction coefficients.

Seating and Unseating Torque, T_s and T_{us}

Seat designs are as numerous as valve manufacturers, and each has its own torque characteristics. This manual presents first-principles methodology for predicting seating and unseating torque through use of seating and unseating coefficients that can be derived from tests for any type of seat. These coefficients may vary with seat material, temperature, and valve pressure rating. Seating torque and unseating torque may be considered to be the same. However, some designs, such as double- or triple-offset designs, may have separate values of coefficients for seating and unseating as well as for each flow direction. The total unseating torque is sometimes referred to as the "break torque."

Closure Member Geometry

Closure member geometry is important to a calculation of valve torque; flowing conditions and cavitation during valve travel (see Figure 3-1). As an example, a symmetrical BFV disc normally has a tendency to close because of the flow passing across the disc. This is similar to the lift of an airplane wing. Most quarter-turn valves normally have a tendency to close at most positions but may have a tendency to open at some positions. This manual does not use the third BFV offset of the triple-offset design in calculations, as it is inconsequential to torque and flow equations and becomes part of the dynamic torque coefficients acquired from testing. The coefficients used in the equations should be specific to and based on the type of valve being evaluated. The offset dimensions ε_1 and ε_2 are shown in Figure 3-1.

Shaft Orientation

When a quarter-turn valve is installed in a horizontal pipe, the shaft orientation is important for calculating torque. When the valve shaft is horizontal in a horizontal pipe with one side empty, the water pressure above and below the shaft is unbalanced and tends to

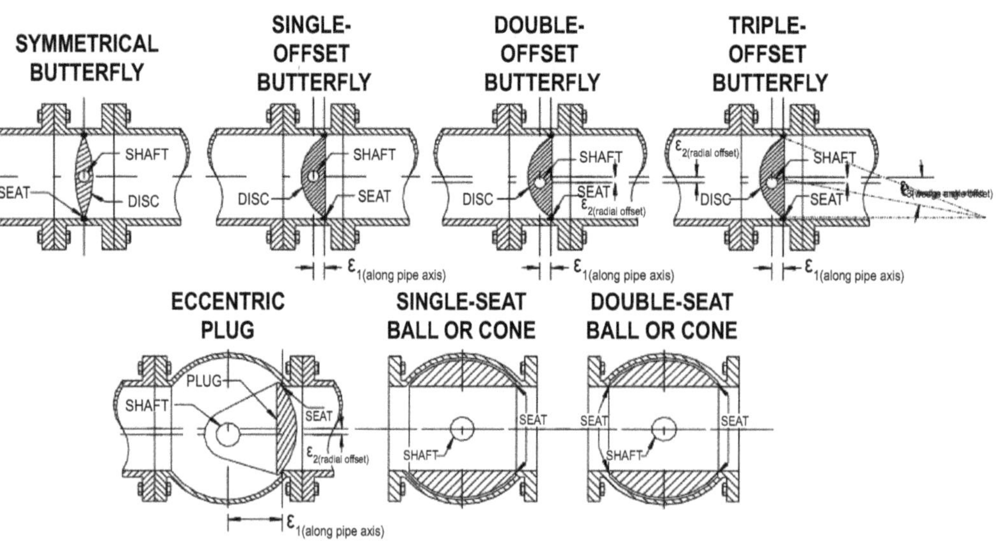

Figure 3-1 Basic closure member design geometries

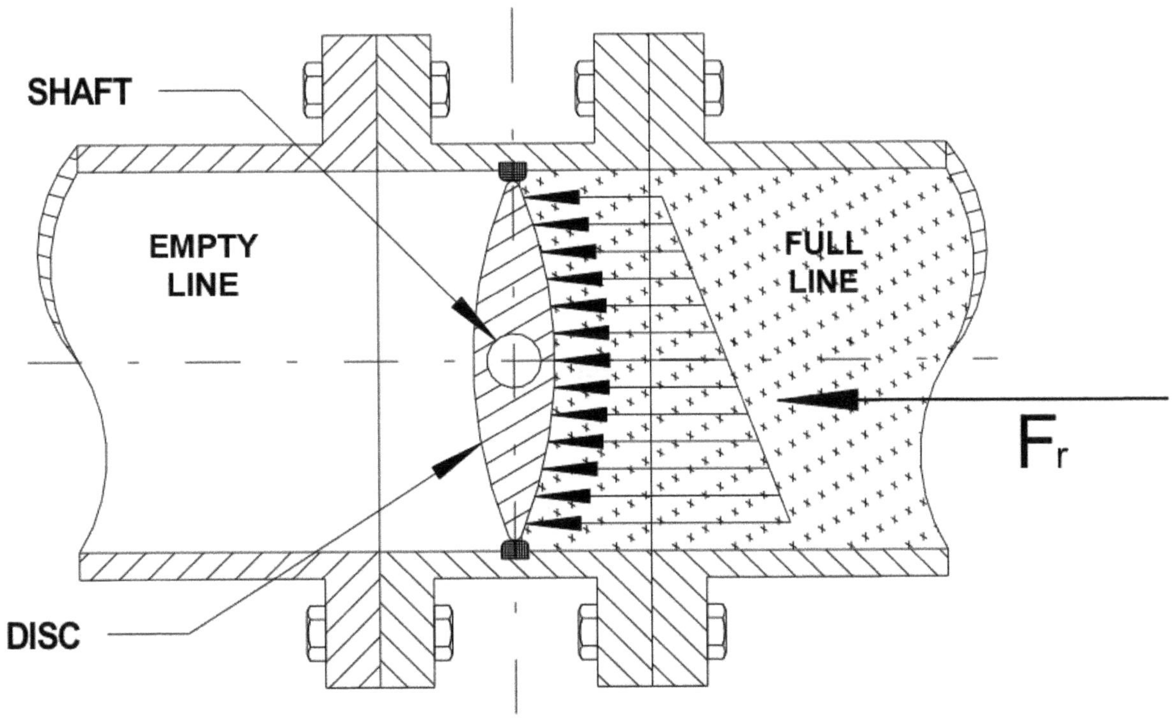

Figure 3-2 Horizontal valve shaft in a horizontal pipe

rotate the valve closure member (see Figure 3-2). This orientation affects the calculations of hydrostatic torque and center-of-gravity torque. For the center-of-gravity torque, the position of the center of gravity when fully open, above or below the shaft axis, determines whether the center-of-gravity torque assists the opening or closing operation.

Flow Direction Through the Valve

Because many quarter-turn valves have offset closure members or other asymmetrical features, the orientation of the valve in the line with respect to flow is important. The valve may have a higher torque requirement with flow toward the shaft side of the closure member or with flow toward the other side. The manufacturer's intended valve orientation must be assured during installation (see Figure 3-3). For purposes of this manual, the flow direction is referred to as seat-side flow or reverse pressure when the seat is upstream of the valve shaft and as shaft-side flow or direct pressure when the seat is downstream of the valve shaft.

Many quarter-turn valves do not have a flow direction preference, but actuator sizing and seat tightness may be dependent on flow direction. Some single-seated BVs and PVs may have a preferred flow direction. Single-seated BVs and PVs used in pump start, stop, and check operation may have installation orientation requirements where the seat is located such as a check valve (between the pump and the valve) to close tight against reverse flow rather than pump flow. In this application, the valve sees reverse pressure when operating open or closed against a running pump and direct pressure when isolating a stopped pump.

Flow in both directions should be considered in the analysis if the installation orientation is not known or if flow reversal is possible.

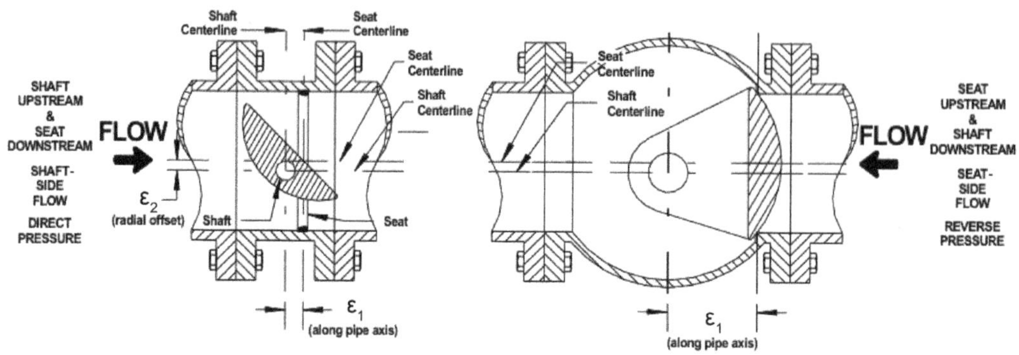

Figure 3-3 Seat-side, reverse, and shaft-side, direct, flow direction orientations

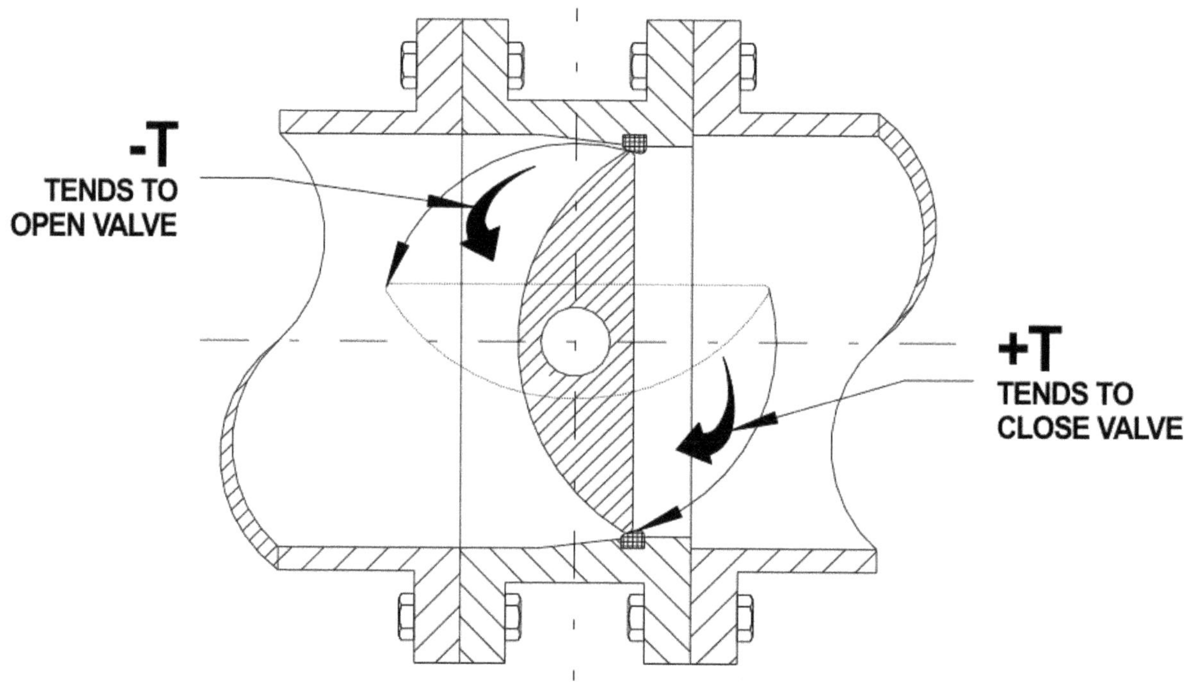

Figure 3-4 Active torque sign convention, positive value tends to close the valve

Torque Sign Conventions

Valve-generated active torque components (T_d, T_h, T_{cg}, and T_{ecc}) are considered as positive values when they tend to close the valve and negative when they tend to open the valve (see Figure 3-4). The signs for friction-based (passive) torque components (T_b, T_s, and T_p) are always considered as positive values because they always oppose actuator motion. Therefore, the total required actuator torque in the opening direction is the summation of all torque components, while the active torque components are subtracted in the closing direction. The most conservative approach for actuator sizing is to sum the absolute values of all torque components, but this may require a substantially oversized actuator.

COMBINING TORQUE COMPONENTS

Quarter-turn actuating valve torque consists of separate elements that contribute into the total valve-actuator torque requirement. The summations of the separate-effects torque calculations given in this section are used to design the valve's shaft and connections as well as select the required actuator size.

This manual is intended to determine the most probable actuator operating torque requirements for the condition analyzed. Data collected and used should be the normal or mean data for the valve and service application. As this method is the summation of many separately generated torque requirements, each torque component will have a possible uncertainty associated with the results. The uncertainty of any torque component may be either positive or negative at any time and normally summed as the square root sum of the squares for the combined total uncertainty. In this manner, the use of application and/or safety factors will not be compounded.

When these calculations and summations are used to select the actuator, they are referred to as the minimum required shaft torque (MRST). The basic actuator's structure is designed with safety margins on the construction material properties and is selected on the basis of its rating and the MRST. Powered actuator sizing is also based on the actuator sizing torque (AST), which includes an application factor (AF) as given in the valve or actuator standard. The AF is chosen on the basis of the valve type, actuator type, and service (on/off or modulating). The MRST is multiplied by an AF to obtain an AST (AST = MRST × AF). This is also calculated at many valve positions to correctly size the actuator. See the valve standards for the AFs to be used. This AST assures that the driving power source has a sufficient strength to fully stroke the valve.

Depending on the position of the closure member, the type of valve installation, and other factor as described here, the general equations used in computing total operating torque requirements or MRST are as follows:

Total seating torque:

$$T_{ts} = T_{b0^\circ} - T_{cg0^\circ} - T_h + T_s + T_p - T_{ecc} \quad \text{(Eq 3-1)}$$

Total unseating (break) torque:

$$T_{tus} = T_{b0^\circ} + T_{cg0^\circ} + T_h + T_{us} + T_p + T_{ecc} \quad \text{(Eq 3-2)}$$

Total opening (run) torque:

$$T_{to\theta} = T_{b\theta} + T_{cg\theta} + T_{d\theta} + T_p \quad \text{(Eq 3-3)}$$

Total closing (run) torque:

$$T_{tc\theta} = T_{b\theta} - T_{cg\theta} - T_{d\theta} + T_p \quad \text{(Eq 3-4)}$$

Where:

Variable	Definition or Description	Units US Customary (SI—Metric)
T_{b0°	Bearing torque at valve angle 0 degrees (always positive)	in.-lb (or ft-lb) (N-m)
$T_{b\theta}$	Bearing torque at valve angle θ (always positive)	in.-lb (or ft-lb) (N-m)

Variable	Definition or Description	Units US Customary (SI—Metric)
$T_{cg\theta}$	Center-of-gravity torque at valve angle θ (Positive value tends to close the valve; negative value tends to open the valve.)	in.-lb (or ft-lb) (N-m)
$T_{cg0°}$	Center-of-gravity torque at valve angle 0 degrees (Positive value tends to close the valve; negative value tends to open the valve.)	in.-lb (or ft-lb) (N-m)
$T_{d\theta}$	Dynamic torque at valve angle θ (Positive value tends to close the valve; negative value tends to open valve.)	in.-lb (or ft-lb) (N-m)
T_{ecc}	Eccentricity torque (Positive value tends to close the valve; negative value tends to open valve.) NOTE: only considered at the seated position during opening or at closing	in.-lb (or ft-lb) (N-m)
T_h	Hydrostatic torque (Positive value tends to close the valve; negative value tends to open the valve.) NOTE: only considered at the seated position during opening or at closing	in.-lb (or ft-lb) (N-m)
T_p	Packing and hub torque (always positive)	in.-lb (or ft-lb) (N-m)
T_s	Seating torque (always positive)	in.-lb (or ft-lb) (N-m)
$T_{tc\theta}$	Total closing torque at valve angle θ (Positive value opposes actuator motion; negative value assists actuator motion.)	in.-lb (or ft-lb) (N-m)
$T_{to\theta}$	Total opening torque at valve angle θ (Positive value opposes actuator motion; negative value assists actuator motion.)	in.-lb (or ft-lb) (N-m)
T_{ts}	Total seating torque (Positive value opposes actuator motion; negative value assists actuator motion.)	in.-lb (or ft-lb) (N-m)
T_{us}	Unseating torque (always positive)	in.-lb (or ft-lb) (N-m)
T_{tus}	Total unseating torque (Positive value opposes actuator motion; negative value assists actuator motion.)	in.-lb (or ft-lb) (N-m)
θ	Subscript indicating valve opening position angle; fully closed equals 0 degrees (0 rad), and fully open normally equals 90 degrees ($\pi/2$ rad) NOTE: Some designs may not travel the full 90 degrees ($\pi/2$ rad) to the full-open position.	degrees (rad)

Historical Note: *Appendix A of AWWA C504, 1987 edition, multiplied the bearing friction torque by a factor of 1.2. This conservatism was removed in the first edition of AWWA M49 to make these calculations more characteristic of actual valve operating conditions and to ensure safety factors and/or allowances for degradation are not duplicated when selecting an actuator. This factor also tended to compensate for the effects of the closure member and shaft weight that were not addressed but are now considered separately in this methodology.*

The total operating torque (T_t) represents the torque or turning effort needed to rotate the closure member. The total torque is usually computed at the closed position (0 degrees, break torque) and at 10 degrees or smaller increments of valve position (run torque). The total torque must be calculated independently for both the opening and closing directions because some torque component signs vary with direction of rotation (see Figures 3-4 and 3-5). Hence, the computed total opening torque at a given angle will be different from the total closing torque at that same angle. Magnitude and direction of torque are essential for selecting actuators that have variable output torque characteristics (such as spring-return cylinder actuators). Actuator sizing recommendations are given in Chapter 6.

Once the magnitude and direction of the component torques are clearly understood, other types of analyses can be performed. For example, a valve and actuator may be sized for the normal maximum system flow rate, but the same assembly must be capable of closing during a line-break flow condition (it will never need to open under line-break flow). The torque calculation at the higher flow rate in the closing direction can be used only to check the actuator size and valve torque capability. Because many valves tend to close as a result of flow, the valve and actuator may be perfectly capable of withstanding a high line-break flow rate with the actuator sized for the normal opening operating flow conditions.

Other special torque calculations include various pump combinations running in a multiple-pump application or reverse-flow conditions (if applicable, as valve dynamic torque often depends on flow direction). For example, when only one pump is running in a multiple-pump application (see Figures 3-6 and 3-7), the pump will run further down on its pump curve and develop a flow rate higher than it would when the other pumps

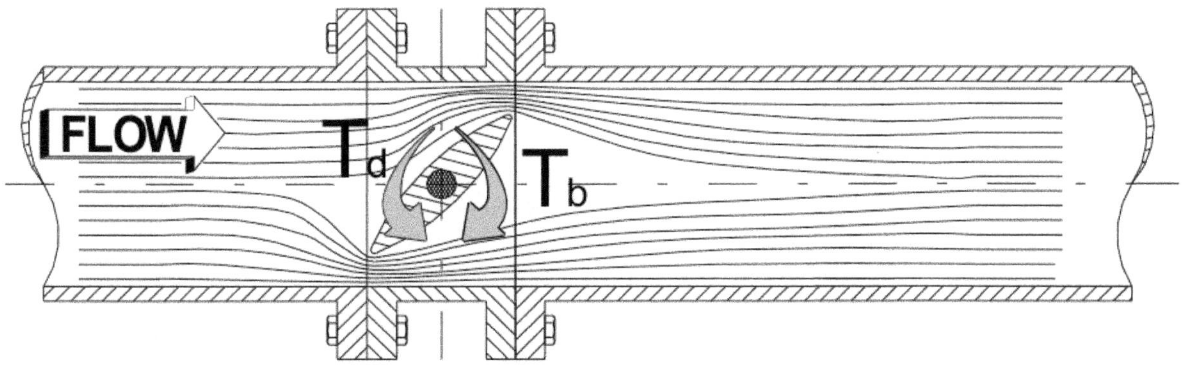

Figure 3-5 Dynamic torque (Td) and bearing torque (Tb) during valve closure

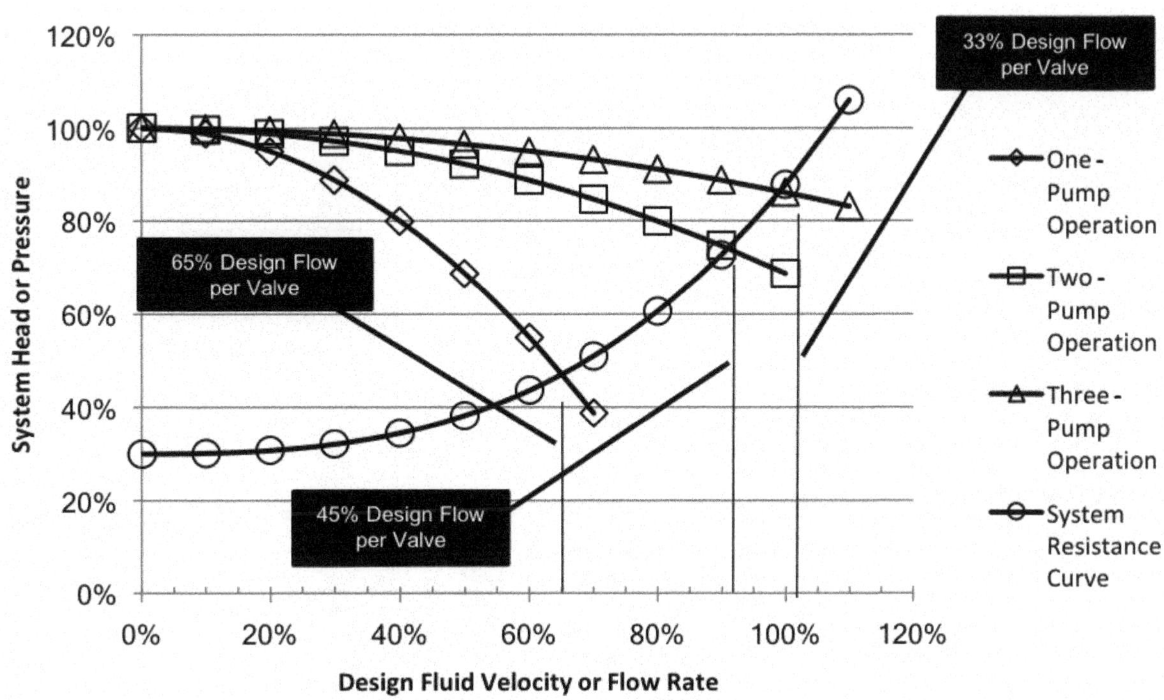

Figure 3-6 One-, two-, and three-pump system operating curves

are operating. This higher flow rate will cause higher valve torques and can stall a power actuator if sized for the flow rate based on all pumps running.

The individual torque components are discussed in detail in the following sections.

SEATING AND UNSEATING TORQUE

In symmetric and single-offset valves, the seating torque (T_s) is caused by the friction and interference between the valve's seat mating surfaces (rubber or metal), as shown in Figure 3-8. In torque-seated double- and triple-offset designs, the seating torque is also based on the seat load necessary to provide the desired level of seat tightness. Seating

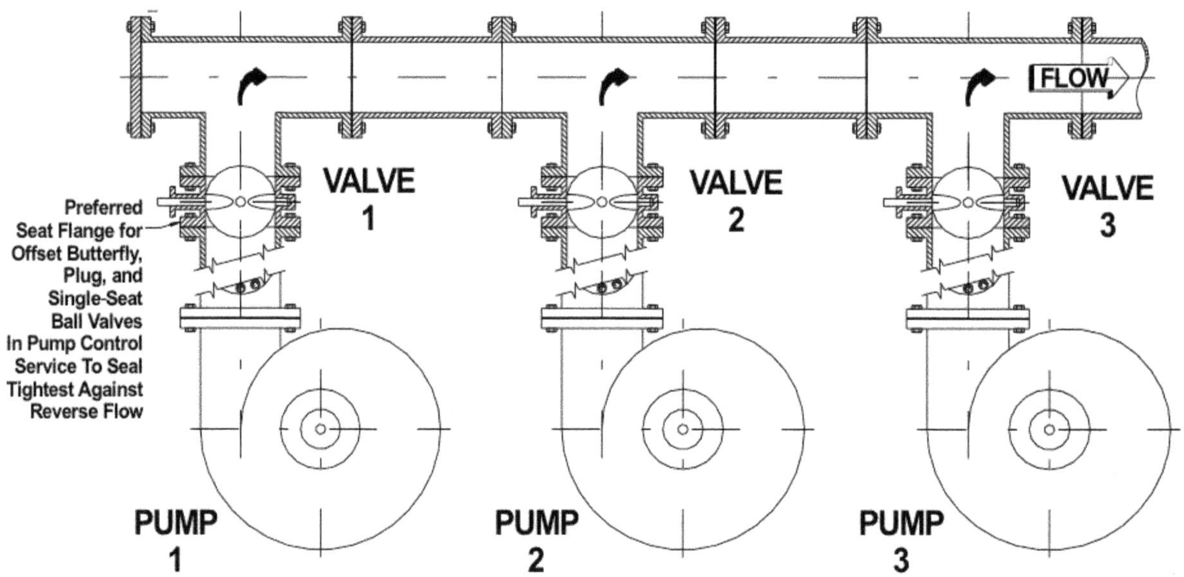

Figure 3-7 Multiple-pump installation, example shown with BFVs

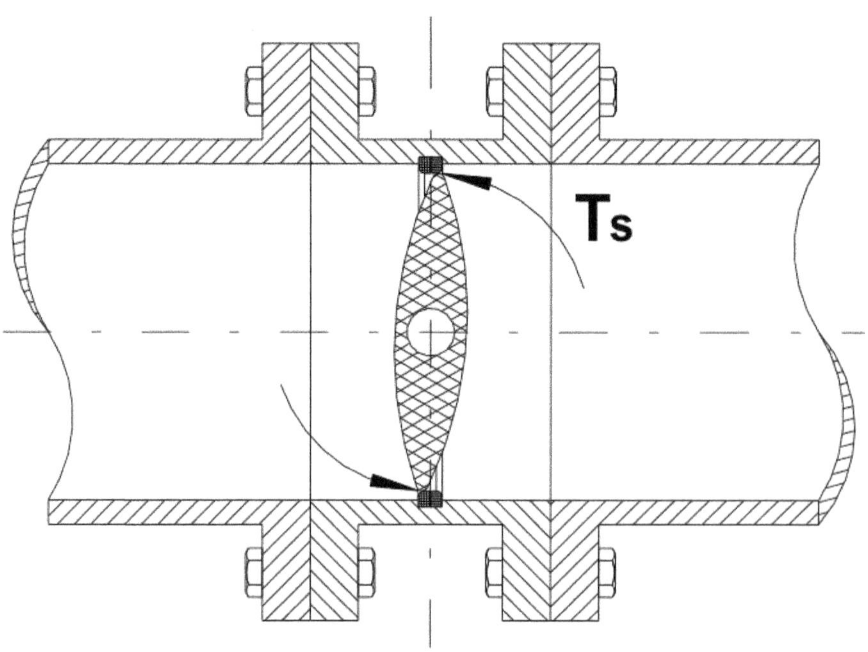

Figure 3-8 BFV seating torque

torque is a function of many factors, including seat type, material, valve size, fluid temperature, and pressure drop across the closure member. The total effect is determined by tests. Given that all other factors are identical, seating torque is normally proportional to the square of the closure member diameter. This formula is derived from first-principles equations and integrated around the perimeter of the seat. This basic derivation is included in Appendix B.

RCVs are a somewhat unique case of a quarter-turn valve in that they are designed to seat and unseat by lifting/dropping the cone (or plug) in a linear stroke direction parallel with the axis of the valve's shaft. A special link/lever mechanism provides for lifting, rotating, and dropping the plug in one smooth linear motion of an input crosshead within the actuator, which provides both torque and lift/thrust. The crosshead may be pushed/pulled by an acme screw stem threaded through it or by a hydraulic cylinder (or hydraulic/pneumatic tandem cylinder). Because of this specialized motion of its closure member, a cone valve's shaft torque can be evaluated only in the midtravel (rotating) portion of its motion. There is no torsional load on the portion of the valve's shaft during the lifting/dropping portion of its motion at the ends of travel. The complete motion of the closure member can be evaluated when the axial resistance to lifting/dropping and the torsional resistance to rotation have each been resolved individually. As this lifting and dropping of the cone involves the thrust calculations of linear valves and the conversion of torque to thrust of a screw stem, the seating and unseating portion of RCVs is not currently covered in this manual.

Many seat designs have essentially constant and pressure-independent friction coefficients, whereas other seat designs are pressure influenced or dependent and friction (load) may increase or decrease with applied pressure. Therefore, two separate seating coefficients are given in the formula. The C_{sc} coefficient is the constant or pressure-independent seating coefficient, and the C_{sp} coefficient is the pressure-dependent seating coefficient. For a given valve design, either of these may be zero. For example, throttling valves that do not have seats do not have a seating torque, and both coefficients are zero. Valves with seats generally have a positive C_{sc} coefficient, but the C_{sp} coefficient may be zero or either a positive or negative value.

BVs and BFVs use Eq 3-5. BVs use D or D_{port} for D_d. BV test data for determining C_{sc} and C_{sp} can be based on solving the same equations using port and/or nominal diameter.

(Ball, butterfly); $$T_s = U_{c2} \times \left(C_{sc} + C_{sp} \times \Delta P_{max}\right) \times D_d^2 \quad \text{(Eq 3-5)}$$

Some manufacturers of double- or triple-eccentric BFVs use the following seat-loading methodology basis for determining the seating torque.

(Double- and triple-offset butterfly);

$$T_s = U_{c2} \times D_d^2 \times W_{seat} \times K \times \Delta P_{max} \times \mu_{seat} \quad \text{(Eq 3-6)}$$

Where:

Variable	Definition or Description	Units US Customary (SI—Metric)
C_{sc}	Constant or pressure-independent coefficient of seating torque*	lb/in. (N/m)
C_{sp}	Pressure-dependent coefficient of seating torque*	lb/in./psi (N/m/kPa)
D_d	Disc diameter (for BVs, use D or D_{port} as D_d)	in. (mm)

Variable	Definition or Description	Units US Customary (SI—Metric)
T_s	Seating torque (always positive)	in.-lb (or ft-lb) (N-m)
U_{c2}	Units' conversion factor: US customary for torque in in.-lb: $U_{c2} = 1$ in./in. US customary for torque in ft-lb: $U_{c2} = 1/12$ (0.0833) ft/in. Metric for torque in N-m: $U_{c2} = 1 \times 10^{-6}$ (0.000001) m^2/mm^2	in./in. (ft/in.) (m^2/mm^2)
ΔP_{max}	Maximum pressure drop (or loss) across the closed valve or total system with valve closed	psid (kPa-d)
W_{seat}	Seat width	in. (mm)
K	Seating compression coefficient, typically equal to 2	none
μ_{seat}	Seat friction coefficient	none

* These variables are dependent on direct or reverse pressure direction application. For torque-seated (double- or triple-offset) valves, the coefficients may be based on the torque required to reach the required pressure and leakage rate.

For PVs, the following equation is used.

$$\text{(Plug);} \qquad T_s = U_{c2} \times \left(C_{sc} + C_{sp} \times \Delta P_{max}\right) \times P_{ml} \times R_p \qquad \text{(Eq 3-7)}$$

Where:

Variable	Definition or Description	Units US Customary (SI—Metric)
P_{ml}	Mean seat perimeter length (see Figure 2-11)	in. (mm)
R_p	Swing radius of PV rubber-covered plug face (see Figure 3-9)	in. (mm)

Seating or unseating torque (T_s or T_{us}) is always positive because it opposes any obturator movement. The effects of seat cleanliness, aging, and degradation are not usually included in the seating coefficient. The test used to determine the seating coefficient (discussed in Chapter 5) is based on a good-condition, well-maintained valve. Manufacturers may apply a safety factor or in-service degradation factor to the seating coefficient or the calculated seating torque to account for long-term service conditions. Wear and aging may decrease or increase the seating torque depending on many valve designs, installation orientations, and service operating conditions; therefore, to apply a degradation factor without good operating experience is inappropriate.

The pressure-dependent coefficient (C_{sp}) represents the change in torque in seat designs that are pressure assisted or otherwise variable on the basis of the operating differential pressure. The pressure-dependent

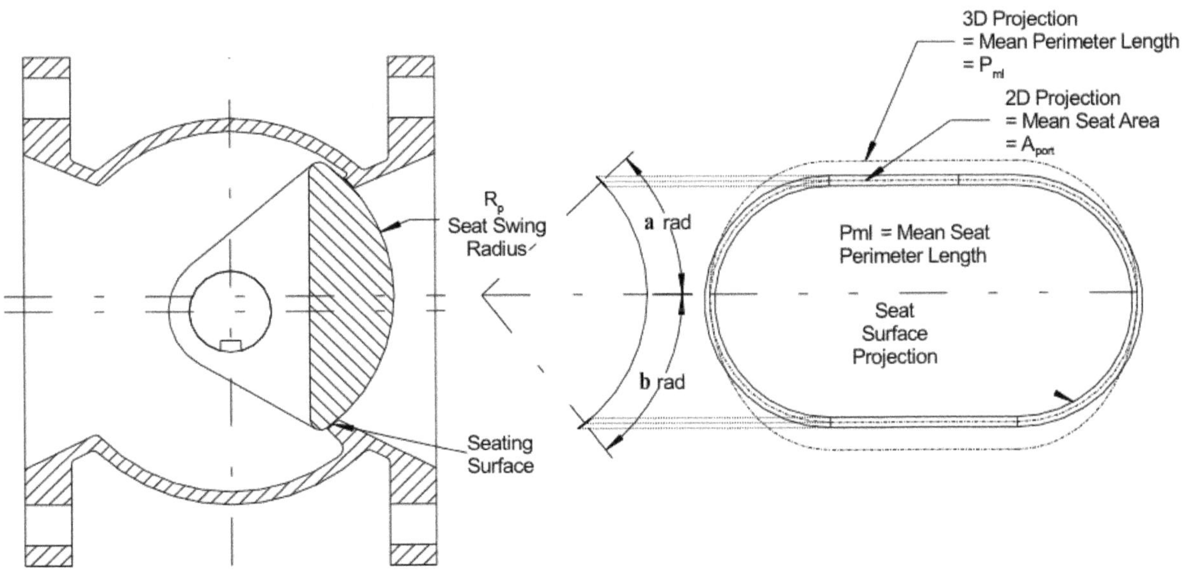

Figure 3-9 PV seat perimeter and port area

coefficient (C_{sp}) may be positive or negative depending on installation orientation. Seat designs that are not affected by pressure differential may have a C_{sp} value of zero.

If the torque required to seat and the torque required to unseat are substantially different, separate seating coefficients and unseating coefficients can be developed in a similar manner using the following equations:

(Ball, butterfly); $$T_{us} = U_{c2} \times \left(C_{usc} + C_{usp} \times \Delta P_{max}\right) \times D_d^2$$ (Eq 3-8)

Where:

Variable	Definition or Description	Units US Customary (SI—Metric)
C_{usc}	Constant or pressure-independent coefficient of seating torque*	lb/in. (N/m)
C_{usp}	Pressure-dependent coefficient of seating torque*	lb/in./psi (N/m/kPa)
D_d	Disc diameter (for BVs, use D or D_{port} as D_d)	in. (mm)
T_s	Seating torque (always positive)	in.-lb (or ft-lb) (N-m)
U_{c2}	Units' conversion factor: US customary for torque in in.-lb: $U_{c2} = 1$ in./in. US customary for torque in ft-lb: $U_{c2} = 1/12$ (0.0833) ft/in. Metric for torque in N-m: $U_{c2} = 1 \times 10^{-6}$ (0.000001) m^2/mm^2	in./in. (in./ft) (m^2/mm^2)
ΔP_{max}	Maximum pressure drop (or loss) across the closed valve or total system with valve closed	psid (kPa-d)

* These variables are dependent on direct or reverse pressure direction application. For torque-seated (double- or triple-offset) valves, the coefficients may be based on the torque required to reach the required pressure and leakage rate.

For PVs:

$$\text{(Plug);} \qquad T_{us} = U_{c2} \times \left(C_{usc} + C_{usp} \times \Delta P_{max}\right) \times P_{ml} \times R_p \qquad \text{(Eq 3-9)}$$

Where:

Variable	Definition or Description	Units US Customary (SI—Metric)
P_{ml}	Mean seat perimeter length	in. (mm)
R_p	Swing radius of PV rubber plug face	in. (mm)

PACKING AND HUB TORQUE

Packing torque (T_P) is caused by friction between the shaft seal (packing) and the valve shaft. The hub seal torque is caused by friction between the closure member and shaft and the body hub seal where the shaft penetrates the pressure boundary (hub) (see Figure 3-10). These are often considered as a single packing torque value. This value is frequently determined by testing and may be given by valve size, shaft diameter, a constant times the shaft diameter, or other formulation.

Packing and hub seal torque (T_P) is always positive because it opposes any closure member movement. This value is usually a small component of total torque and is often

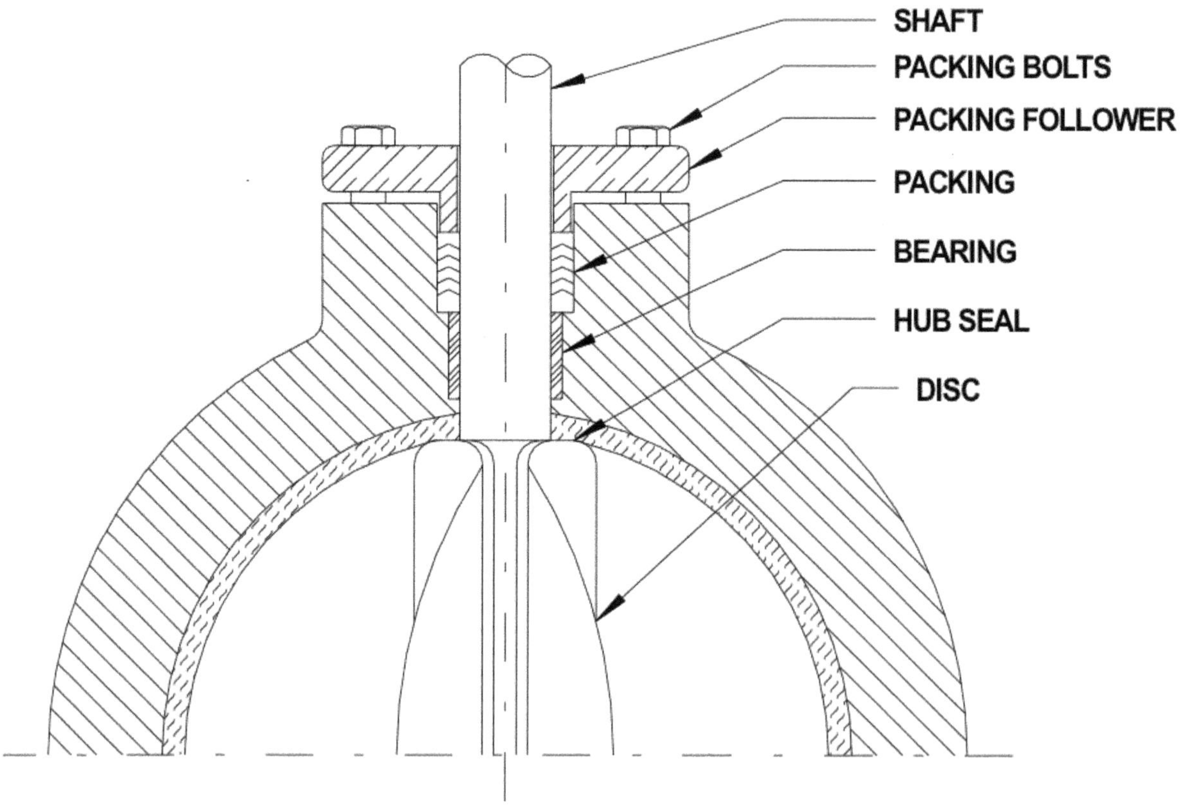

Figure 3-10 BFV packing and hub seal torque (Tp)

ignored on larger valves. When the shaft seal is of the O-ring type or V-packing type, this component of torque is not significant and may be ignored. However, overtightening of the shaft packing bolts or studs can cause a significant packing torque increase. Consult the valve manufacturer for packing adjustment instructions and recommendations. In some cases, this torque may be considered as a component of seating torque or of other frictional components of torque and may be assumed to be zero.

As packing is friction-based and there are many types of packing, loading methods, and adjustment procedures, there is a broad variation in friction data. Many packing manufacturers provide packing friction ranges and calculation procedures based on packing type, material, method of loading, and operating pressure. Some typical packing torque calculations that may be used are:

$$T_p = U_{c1} \times C_{pckf} \times d_s^2 \qquad \text{(Eq 3-10)}$$

$$T_p = U_{c1} \times C_{pck\ q} \times d_s \qquad \text{(Eq 3-11)}$$

or for chevron-type packing:

$$T_p = U_{c1} \times \frac{3 \times pc \times \pi \times \upsilon \times H_p \times \mu_p}{4} \times d_s^2 \qquad \text{(Eq 3-12)}$$

Where:

Variable	Definition or Description	Units US Customary (SI—Metric)
$C_{pck\ f}$	Packing force coefficient	lb/in. (N/m)
$C_{pck\ t}$	Packing torque coefficient	lb (N)
D_s	Shaft diameter	in. (m)
H_p	Packing height	in. (m)
P_c	Pressure class or maximum design pressure (the greater of)	psi (k/Pa)
T_p	Packing and hub seal torque (always positive)	in.-lb (or ft-lb) (N-m)
U_{c1}	Units' conversion factor: US customary for torque in in.-lb: U_{c1} = 1 in./in. US customary for torque in ft-lb: U_{c1} = 1⁄12 (0.0833) in./ft Metric for torque in N-m: $U_{c1} = 1 \times 10^{-3}$ (0.001) m/mm	in./in. (in./ft) (m/mm)
μ_P	Packing coefficient of friction (typically 0.1–0.3)	dimensionless
ν	Packing radial stress to axial stress transfer ratio (typically assumed to be ≈0.5)	dimensionless

BEARING TORQUE

The bearing torque (T_b) in a quarter-turn valve is a function of the coefficient of friction between the bearing and the shaft or trunnion, the shaft or trunnion diameter, the closure member diameter (or area), combined closure member and shaft(s) weight, vertical orientation of the shaft axis, and the pressure drop across the closure member at each angle of rotation (see Figure 3-11).

The differential pressure for these calculations is either the calculated differential when unchoked or the pressure just upstream minus the vapor pressure of the water when a choked condition exists.

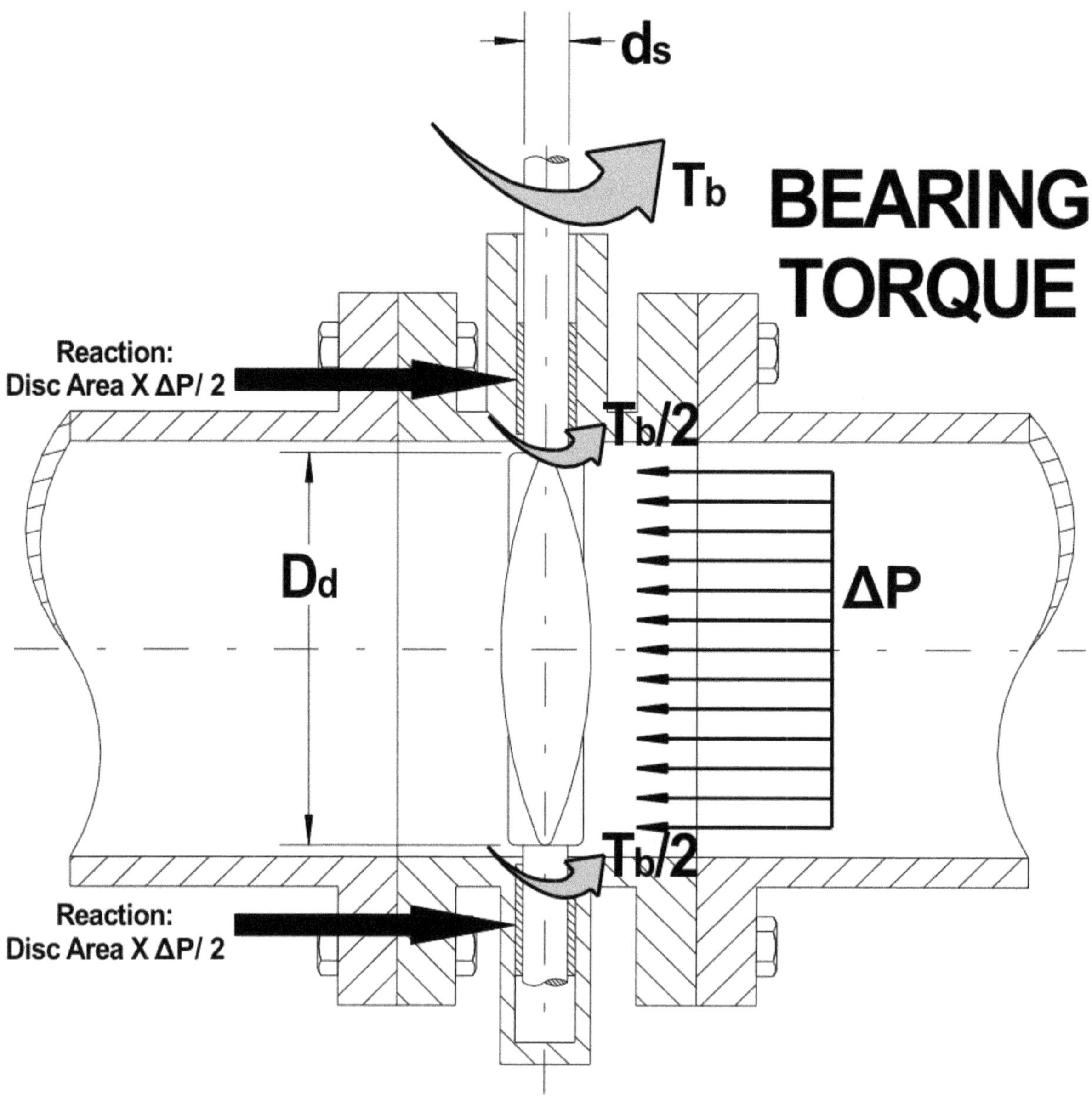

Figure 3-11 Bearing friction torque (T_b)

Ball, Butterfly, and Rotary Cone Valves:

For bearing torque due to differential pressure only:

$$T_{b\theta} = U_{c2} \times \frac{\pi \times D_d^2 \times \Delta P_\theta \times d_s \times C_f}{8} \qquad \text{(Eq 3-13)}$$

For bearing torque due to differential pressure plus closure member and shaft weight as a conservative direct summation†:

$$T_{b\theta} = U_{c2} \times \frac{\left(\pi \times D_d^2 \times \Delta P_{brg\theta} + W_{d\&s}\right) \times d_s \times C_f}{8} \qquad \text{(Eq 3-14)}$$

or:

$$P_{ew} = \frac{4 \times W_{d\&s}}{\pi \times D_d^2} \qquad \text{(Eq 3-15)}$$

and:

$$T_{b\theta} = U_{c2} \times \frac{\pi \times D_d^2 \times \left(\Delta P_{brg\theta} + P_{ew}\right) \times d_s \times C_f}{8} \qquad \text{(Eq 3-16)}$$

Where:

Variable	Definition or Description	Units US Customary (SI—Metric)
C_f	Coefficient of friction between the shaft or trunnion and bushing, dimensionless (This value may be obtained from a flow test, engineering handbooks, the bearing manufacturer, or the valve manufacturer.)	dimensionless
D_d	Disc diameter for BFVs or nominal diameter (D) or port diameter (D_{port}) for BVs and RCVs	in. (mm)
d_s	Shaft diameter. For trunnion-mounted valves, use trunnion diameter (d_{tr}).	in. (mm)
d_{tr}	Trunnion diameter for trunnion-mounted obturators	in. (mm)
P_{ew}	Pressure equivalent to closure member and shaft weight	psi (kPa)
$T_{b\theta}$	Bearing torque at valve angle θ (always positive)	in.-lb (or ft-lb) (N-m)
$W_{d\&s}$	Weight of the obturator (ball, disc, or plug) and shaft(s) assembly (banjo). For applications involving additional seismic loading, $W_{d\&s}$ may be multiplied by G or (G±1), where G is the additional gravitational acceleration multiplier.	lb (kg)

Variable	Definition or Description	Units US Customary (SI—Metric)
$\Delta P_{brg\theta}$	Pressure drop (or loss), ΔP_θ, while at closure member angle θ during classical flow or $P_{ua\theta} - P_{va\theta}$ during choked flow	psid (kPa-d)
P_{ua}, $P_{ua\theta}$	Reference pressure just upstream of the valve for choking and cavitation analysis. Subscript θ indicates the valve angle.	psia (kPa)
P_{va}, $P_{va\theta}$	Absolute vapor pressure (Examples: P_{va} = 0.0867 psia [0.597 kPa] for water at 32°F [0°C] and P_{va} = 1.93 psia [13.3 kPa] for water at 125°F [51.7°C] measured at sea level). Subscript θ indicates the valve angle.	psia (kPa)
ΔP_θ	Pressure drop (or loss) while at closure member angle θ	psid (kPa-d)

Plug Valves

For bearing torque due to differential pressure only:

$$T_{b\theta} = U_{c2} \times \frac{A_{port} \times \Delta P_{brg\theta} \times d_{sm} \times C_f}{2} \qquad \text{(Eq 3-17)}$$

For bearing torque due to differential pressure plus closure member and shaft weight as a conservative direct summation†:

$$T_{b\theta} = U_{c2} \times \frac{\left(A_{port} \times \Delta P_{brg\theta} + W_{d\&s}\right) \times d_{sm} \times C_f}{2} \qquad \text{(Eq 3-18)}$$

or:

$$P_{ew} = \frac{W_{d\&s}}{A_{port}} \qquad \text{(Eq 3-19)}$$

and:

$$T_{b\theta} = U_{c2} \times \frac{A_{port} \times \left(\Delta P_\theta + P_{ew}\right) \times d_{sm} \times C_f}{2} \qquad \text{(Eq 3-20)}$$

Where:

Variable	Definition or Description	Units US Customary (SI—Metric)
C_f	Coefficient of friction between the shaft or trunnion and bushing, dimensionless (This value may be obtained from a flow test, engineering handbooks, the bearing manufacturer, or the valve manufacturer.)	dimensionless
A_{port}	Valve mean port area based on the center or mean seating point around the perimeter of the seat	$in.^2$ (mm^2)

Variable	Definition or Description	Units US Customary (SI—Metric)
d_{sm}	Mean shaft diameter. Top or actuator shaft and the bottom or blind shaft may be different diameters.	in. (mm)
P_{ew}	Pressure equivalent to obturator and shaft weight*	psi (kPa)
$T_{b\theta}$	Bearing torque at valve angle θ (always positive)	in.-lb (or ft-lb) (N-m)
$W_{d\&s}$	Weight of the obturator (ball, disc, or plug) and shaft(s) assembly (banjo). For applications involving additional seismic loading, $W_{d\&s}$ may be multiplied by G or (G±1), where G is the additional gravitational acceleration multiplier.*	lb (kg)
$\Delta P_{brg\theta}$	Pressure drop (or loss), ΔP_θ, while at closure member angle θ during classical flow or $P_{ua\theta} - P_{va\theta}$ during choked flow	psid (kPa-d)
P_{ua}, $P_{ua\theta}$	Reference pressure just upstream of the valve for choking and cavitation analysis. Subscript θ indicates the valve angle.	psia (kPa)
P_{va}, $P_{va\theta}$	Absolute vapor pressure (Examples: P_{va} = 0.0867 psia [0.597 kPa] for water at 32°F [0°C] and P_{va} = 1.93 psia [13.3 kPa] for water at 125°F [51.7°C] measured at sea level). Subscript θ indicates the valve angle.	psia (kPa)
ΔP_θ	Pressure drop (or loss) while at plug angle θ	psid (kPa-d)

* Eqs 3-18 through 3-20 conservatively add the weight of the closure member and shafts directly in the bearing friction calculation. This is a worst-case approach, as the bearing force load caused by the differential pressure and the force caused by the closure member and shaft(s) weight should be added as vector sums with respect to the installed orientations. The addition of this weight component to this torque is normally insignificant at higher differential pressures and is significant only at low differential pressures or when high seismic loads are required. This generally adds between 0.5 and 4.0 psi (3.45 and 27.6 kPa) to the calculated differential pressure for valves within the scope of this manual. See the section in this chapter titled "Other Components of Torque" for a more complete methodology for critical applications. For applications involving additional seismic loading, $W_{d\&s}$ may be multiplied by G (horizontal) or G±1 (vertical), where G is the additional seismic acceleration.

Bearing torque is always positive because it opposes any closure member movement. The value is highest at the near-closed position because of the high differential pressure when the valve is nearly closed. The bearing torque reduces to almost zero as the valve reaches the fully open position.

Bearing Load Lift and Drag Rectangular Coordinates

Some further research (e.g., G.A. Schohl [2004], CFD Prediction and validation of Butterfly Valve Hydrodynamic Forces) shows that the hydrodynamic forces around the disc can be evaluated into both lift and drag components. Although this adds to the understanding of the bearing loads, this also adds complexity and is only necessary to resolve these loads into the rectangular coordinates (see Figure 3-12).

CENTER-OF-GRAVITY TORQUE

Center-of-gravity torque (T_{cg}) is caused by the offset center of gravity of the closure member. This torque occurs when the valve shaft is located in or near the horizontal plane, and

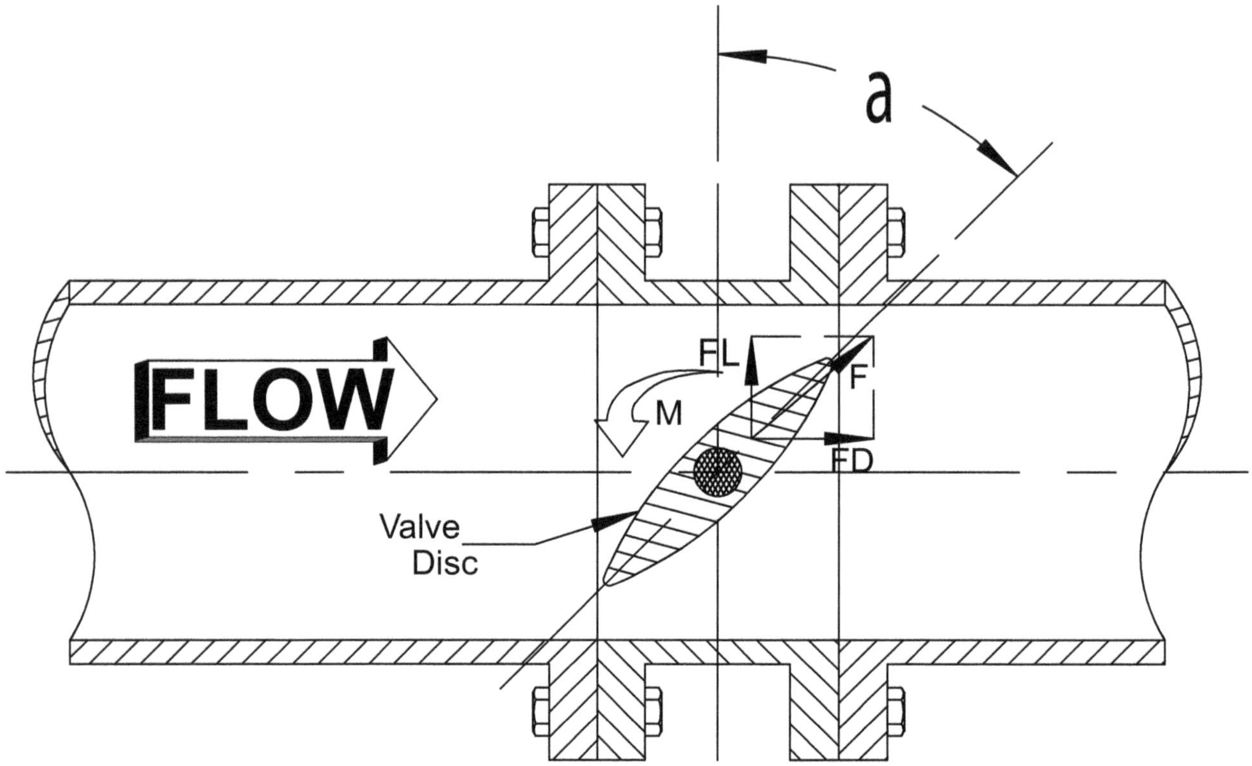

Figure 3-12 Hydrostatic force on a partially open BFV

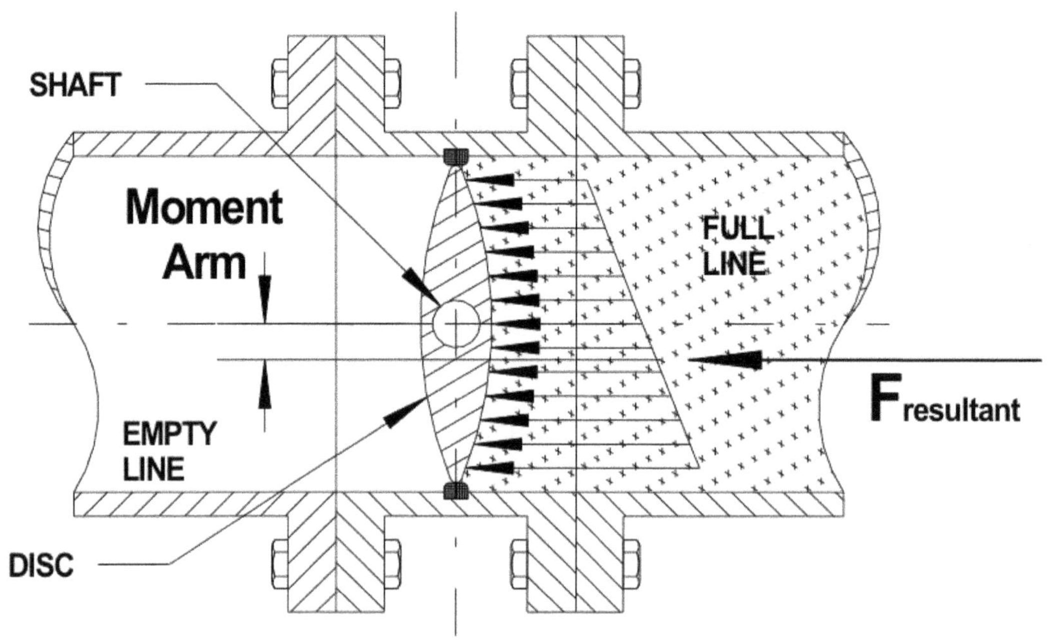

Figure 3-13 Center-of-gravity torque (T_{cg})

it is a function of the closure member position and weight as well as the distance from the axis of rotation to the center of gravity (see Figure 3-13). With a horizontal valve shaft and a horizontal pipeline, this torque is greatest when the center-of-gravity location is situated on the pipeline axis. This torque varies considerably on the basis of closure member design and the center-of-gravity location. This torque may be assumed as insignificant or may be included as a worst-case constant maximum value throughout the valve travel by setting COS(θ + γ) equal to one (1) throughout the valve travel. See the section on Other Torque Components and Figures 3-24 and 3-25 for more complete installation details for use in critical applications in which the pipe axis and shaft axis are known to be in other orientations. BVs and RCVs have zero (0) or a very small C_g, and this torque is generally not applicable to these valves.

For horizontal shaft and pipe axis, the basic equation is:

$$T_{cg\theta} = U_{c1} \times S_c \times W_d \times C_g \times \text{COS}(\theta + \gamma) \quad \text{(Eq 3-21)}$$

Where:

Variable	Definition or Description	Units US Customary (SI—Metric)
C_g	Valve closure member center-of-gravity distance from shaft centerline	in. (mm)
S_c	Sign convention variable: For torque active components, +1 when the torque tends to close the valve or –1 when the torque tends to open the valve For center-of-gravity torque, the sign convention variable is +1 when the center of gravity is above the horizontal when the closure member is in the full-open position or –1 when the center of gravity is below the horizontal when closure member is full open. For torque transmitted to the actuator: a positive value when valve shaft torque opposes actuator motion or negative value when valve shaft torque assists actuator motion (actuator is acting as a brake)	dimensionless
$T_{cg\theta}$	Center-of-gravity torque at valve angle θ (Positive value tends to close the valve; negative value tends to open the valve.)	in.-lb (or ft-lb) (N-m)
W_d	Weight of valve closure member	lb (kg)
γ	Center-of-gravity offset angle (nonsymmetric closure member designs)	degrees (rad)
θ	Valve opening position angle; closed = 0 degrees (0 rad), and fully open ≈ 90 degrees ($\pi/2$ rad)	degrees (rad)

HYDROSTATIC TORQUE

Hydrostatic torque (T_h) is caused by the static elevation head of the water acting against one side of the closure member when the other side of the closure member is empty (refer to Figures 3-14 and 3-26). This torque component occurs when pipe axis and shaft axis are at or near horizontal (pipe and shaft angles greater than zero). The sign convention variable may be assigned to indicate when assisting or opposing actuator motion, or it might be considered to oppose both operating directions as a worst-case assumption. This torque component only occurs when there is no flow and the valve is seated and one side of the line is empty.

When the water pressure forces on the closure member are integrated across the circular surface, it can be replaced with a single force vector, $F_{resultant}$, as shown in Figure 3-14. Although not used in the derivation of the formula, this vector provides a clearer visualization of the effect the fluid pressure load. The derivation of this torque is not included but is based on a circular area. Although PVs may have a noncircular port area, this derivation is suitable for use with all valves within this standard practice.

When valve shaft and pipe axis are horizontal:

(US customary units); $$T_h = S_c \times U_{c1} \times \frac{\rho \times \pi}{5.333} \times \left(\frac{D_d}{12}\right)^4 \times \left(1 + \frac{8 \times \varepsilon 2}{D_d}\right) \quad \text{(Eq 3-22)}$$

or:

(SI metric units); $$T_h = S_c \times \frac{\rho \times \pi \times g}{64} \times (U_{c1} \times D_d)^4 \times \left(1 + \frac{8 \times \varepsilon_2}{D_d}\right) \quad \text{(Eq 3-23)}$$

or:

For cold water where $\rho = 62.4$ lb/ft^3:

(US customary units); $$T_h = S_c \times U_{c1} \times 0.0017726 \times (D_d)^4 \times \left(1 + \frac{8 \times \varepsilon_2}{D_d}\right) \quad \text{(Eq 3-24)}$$

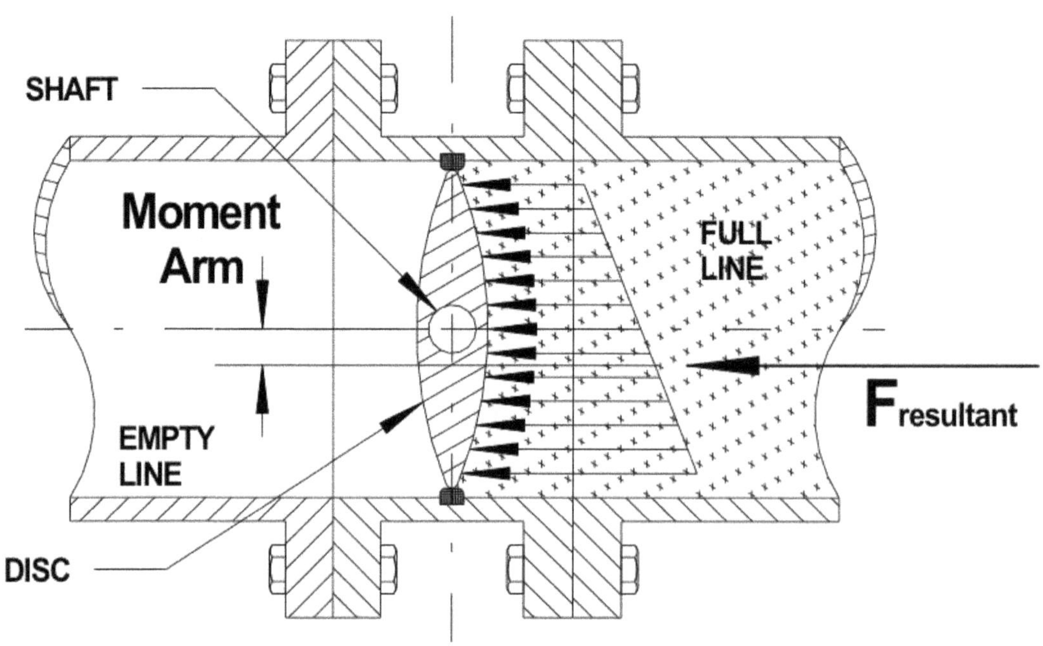

Figure 3-14 Hydrostatic torque (T_h)

For cold water where $\rho = 1{,}000$ kg/m^3:

(SI metric units);

$$T_h = S_c \times 481.322 \times (U_{c1} \times D_d)^4 \times \left(1 + \frac{8 \times \varepsilon_2}{D_d}\right) \quad \text{(Eq 3-25)}$$

Where:

Variable	Definition or Description	Units US Customary (SI—Metric)
D_d	Closure member diameter; for noncircular (PV) ports, use $D_d = (4 \times A_{port}/\pi)^{1/2}$	in. (mm)
g	Gravitational constant Acceleration due to gravity, 32.2 ft/s^2 (9.81 m/s^2)	ft/s^2 (m/s^2)
S_c	Sign convention variable: For torque active components, +1 when the torque tends to close the valve or –1 when the torque tends to open the valve For center-of-gravity torque, the sign convention variable is +1 when the center of gravity is above the horizontal when the closure member is in the full-open position or –1 when the center of gravity is below the horizontal when closure member is in the full-open position. For torque transmitted to the actuator: a positive value when valve shaft torque opposes actuator motion or negative value when valve shaft torque assists actuator motion (actuator is acting as a brake)	dimensionless
T_h	Hydrostatic torque (Positive value tends to close the valve; negative value tends to open the valve.) NOTE: only considered at the seated position during opening or at closing	in.-lb (or ft-lb) (N-m)
U_{C1}	Units' conversion factor: US customary for torque in in.-lb: U_{c1} = 1 in./in. US customary for torque in ft-lb: U_{c1} = 1⁄12 (0.0833) ft/in. Metric for torque in N-m: $U_{c1} = 1 \times 10^{-3}$ (0.001) m/mm	in./in. (ft/in.) (m/mm)
ε_2	Closure member lateral offset NOTE: ε_2 equals 0 for symmetric or single-offset closure member designs. Sign convention note: For hydrostatic torque, ε_2 is positive when oriented above the valve centerline and negative when oriented below the valve centerline. See Figure 3-3.	in. (mm)
ρ	Fluid density NOTE: Standard water density is considered as 62.4 lb/ft^3 (1,000 kg/m^3) at 40°F (4.4°C). Also see Table 4 in the Nomenclature section. Refer to Crane (2009 or later), Flow of Fluids Through Valves, Fittings and Pipe, Technical Paper No. 410	lb/ft^3 (kg/m^3)

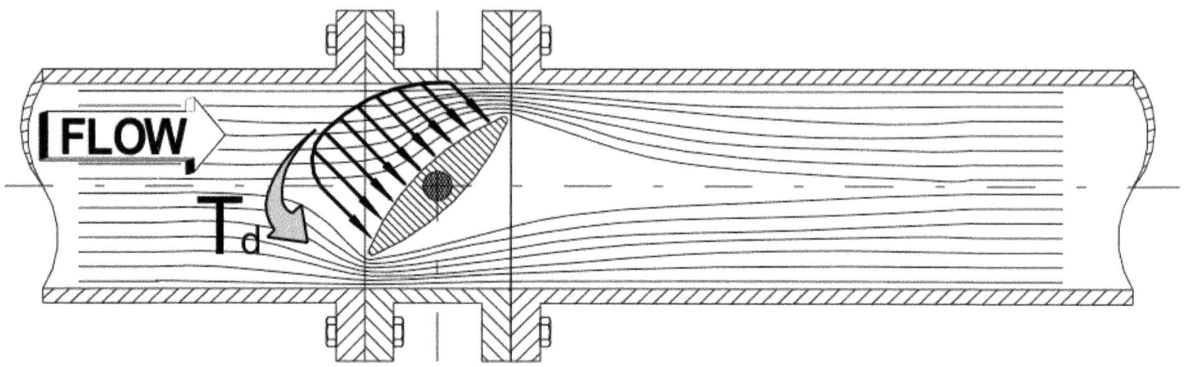

Figure 3-15 Dynamic torque (T_d) for a symmetrical disc BFV

DYNAMIC TORQUE

Dynamic torque (T_d) is a flow-induced torque determined as a function of valve geometry, flow rate, and valve position (Figure 3-15).

The differential pressure for these calculations is either the calculated differential when unchoked (classical flow) or the choked differential pressure, $\Delta P_{choke\theta}$, when the flow is choked at the valve.

$$T_{d\theta} = U_{c2} \times C_{t\theta} \times D_d^3 \times \Delta P_{dyn\theta} \quad \text{(Eq 3-26)}$$

Where:

Variable	Definition or Description	Units US Customary (SI—Metric)
$C_{t\theta}$	Coefficient of dynamic torque at valve angle θ (Positive value tends to close valve.)	dimensionless
D_d	Disc diameter*	in. (mm)
$T_{d\theta}$	Dynamic torque at valve angle θ (Positive value tends to close the valve; negative value tends to open valve.) NOTE: There are several other names used in conjunction with this torque component such as fluidynamic torque, hydrodynamic torque, or aerodynamic torque.	in.-lb (or ft-lb) (N-m)
U_{c2}	Units' conversion factor: US customary for torque in in.-lb: U_{c2} = 1 in./in. US customary for torque in ft-lb: U_{c2} = 1⁄12 (0.0833) ft/in. Metric for torque in N-m: $U_{c2} = 1 \times 10^{-6}$ (0.000001) m^2/mm^2	in./in. (ft/in.) (m^2/mm^2)
$\Delta P_{dyn\theta}$	Pressure drop (or loss), ΔP_θ, while at closure member angle θ during classic flow or $\Delta P_{choke\theta}$ during choked flow	Psid (kPa)
ΔP_θ	Pressure drop (or loss) while at closure member angle θ	psid (kPa)

* The nominal valve diameter (D) will be used for BVs, RCVs, and PVs. The nominal valve diameter (D) is often used for BFVs when the D_d is unknown, which tends to increase the uncertainty but produces conservatively larger torque values.

The maximum coefficient of dynamic torque (which reflects a constant pressure drop at all travel positions) occurs at approximately the 65–80 degree open position. In contrast, the maximum total dynamic torque (the summation of all operating torque requirements) normally occurs at an intermediate travel position between 0 (closed) and 50 degrees (open) where the differential pressure is high (that is, pressure drop varies with valve position). Pressure drop and dynamic torque are dependent on the characteristics of the piping system and cannot be assumed without a system analysis.

Dynamic torque coefficients for a symmetrical closure member are normally independent of flow direction (Figure 3-16). They are functions of closure member geometry, valve travel, and pressure drop.

Dynamic torque coefficients for a single-offset BFV disc, as shown in Figure 3-16, depend on flow direction through the valve as well as disc geometry, valve travel, and pressure drop across the BFV disc. The dynamic torque coefficient at the open position may be negative (giving the closure member a tendency to open) when the valve is installed with the seat upstream. The BFV single-offset disc torque coefficient can also change sign near the 85-degree position. If the valve is positioned at an angle at which the torque direction is unstable (when C_t crosses zero at angles other than 90 degrees) for extended periods, fatigue damage caused by torque reversals and vibration may occur, and prolonged modulation in this valve position should be avoided.

Although the dynamic torque coefficients reach maximum at about 70–80 degrees (78–90 percent) open (as shown in Figure 3-16), the total valve opening torque reaches maximum at a much lower angle (about 35 degrees or 39 percent) as shown in Figures 3-18

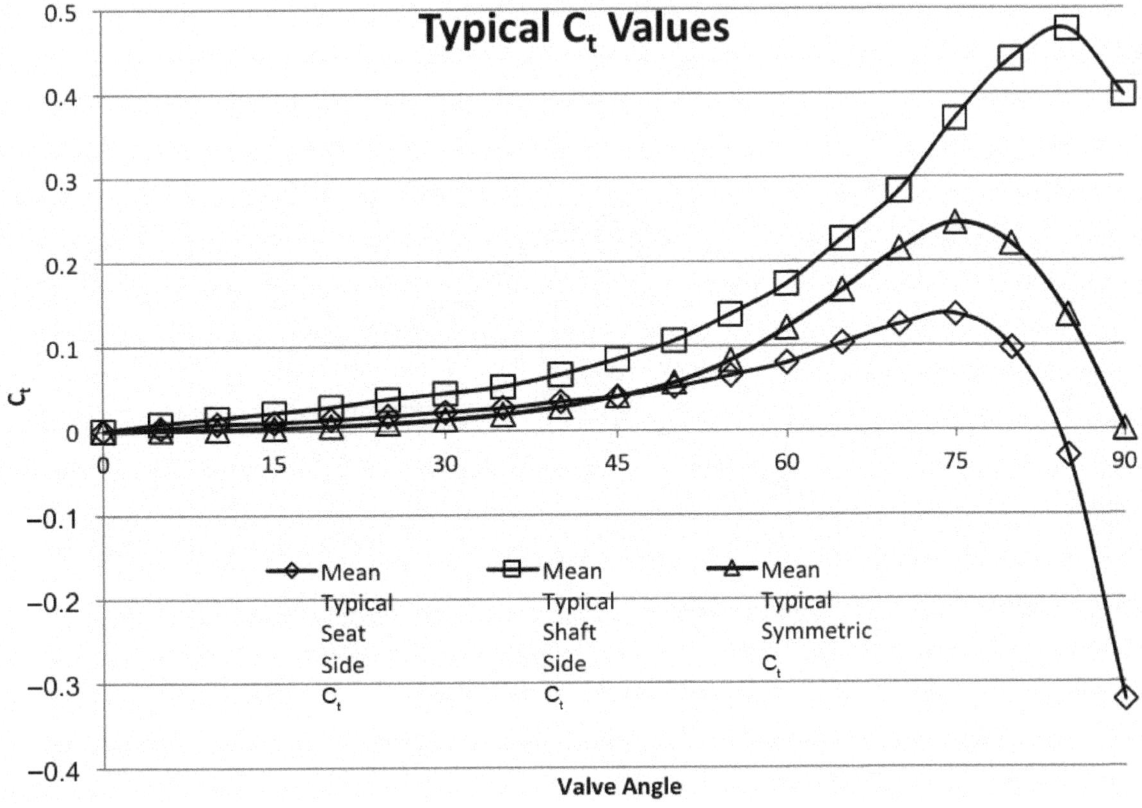

Figure 3-16 Dynamic torque coefficient (C_t) graph for BFVs with symmetrical and offset closure members

and 3-19. Figures 3-20 and 3-21 demonstrate how the total actuator torque changes during the closing stroke. In Figure 3-17, dynamic torque is highest at ≈45 degrees (50 percent) open because the pressure drop (ΔP) is an order of magnitude higher there than at 80 degrees (90 percent) open, and dynamic torque is the product of both the dynamic torque coefficient and the pressure drop (see Eq 3-26).

Because dynamic torque is proportional to the closure member diameter cubed, it often becomes the largest torque on valves greater than about 20 in. (500 mm) for velocities near 16 ft/s (5.2 m/s) and above. On smaller valves, typically 6 in. (150 mm) and less, dynamic torque can be ignored, and the actuator may be sized for seating, bearing, and packing torque unless the maximum velocity exceeds 16 ft/s (5.2 m/s).

Because the dynamic torque is dependent on the pressure drop during the valve travel, the dynamic torque coefficient and the flow coefficients must be from the same valve data set and not mixed with coefficients from other valves.

SHAFT OFFSET OR ECCENTRICITY TORQUE

Shaft offset or eccentricity torque for a quarter-turn valve with a double- or triple-offset shaft is shown in Figure 3-22. This design is subject to an additional torque related to the lateral offset or eccentricity and the hydrostatic differential force on the closure member. This torque only applies at the full-closed or seated position as it becomes a part of the dynamic torque coefficients and calculations at other closure member positions. This torque is often treated as a positive value but may be given a sign if the seating and unseating torque components are to be considered individually.

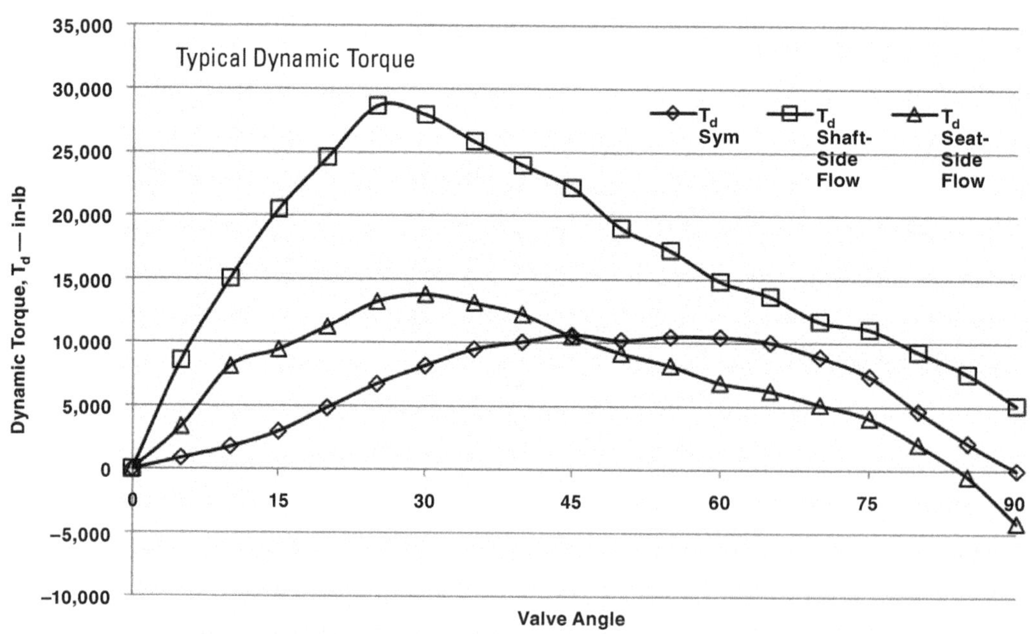

Figure 3-17 Dynamic torque (T_d) for a BFV with symmetric and offset discs (NOTE: Metric torque is 565 N-m per division)

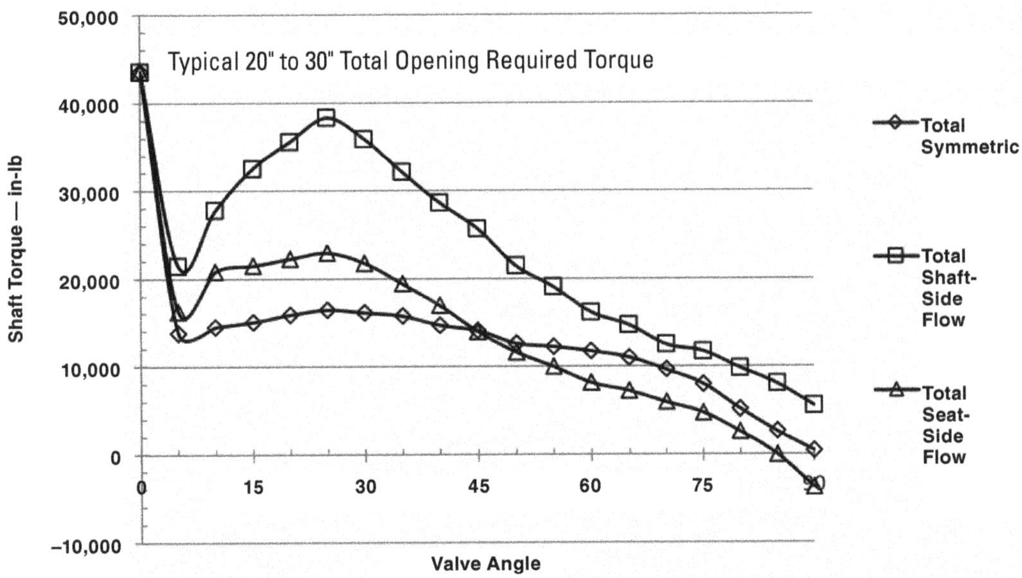

Figure 3-18 **Typical total opening torque (T_{to}) for a 20-in. (500-mm) to 30-in. (750-mm) BFV with symmetric and offset discs in each flow direction (NOTE: Metric torque is 1,130 N-m per division)**

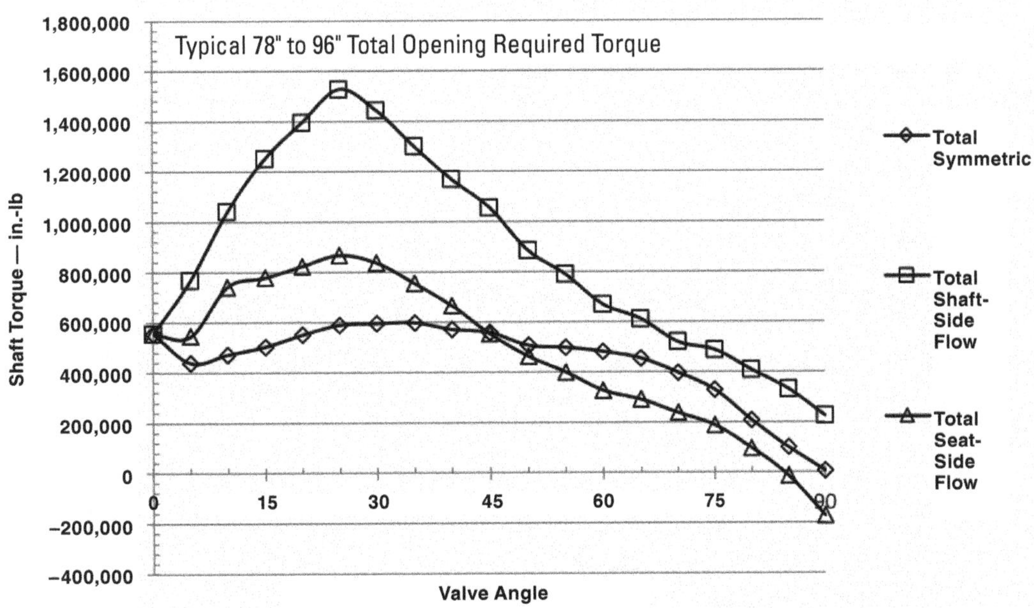

Figure 3-19 **Typical total opening torque (T_{to}) for a 78-in. (1,950-mm) to 96-in. (2,400-mm) BFV with symmetric and offset discs in each flow direction (NOTE: Metric torque is 22,600 N-m per division)**

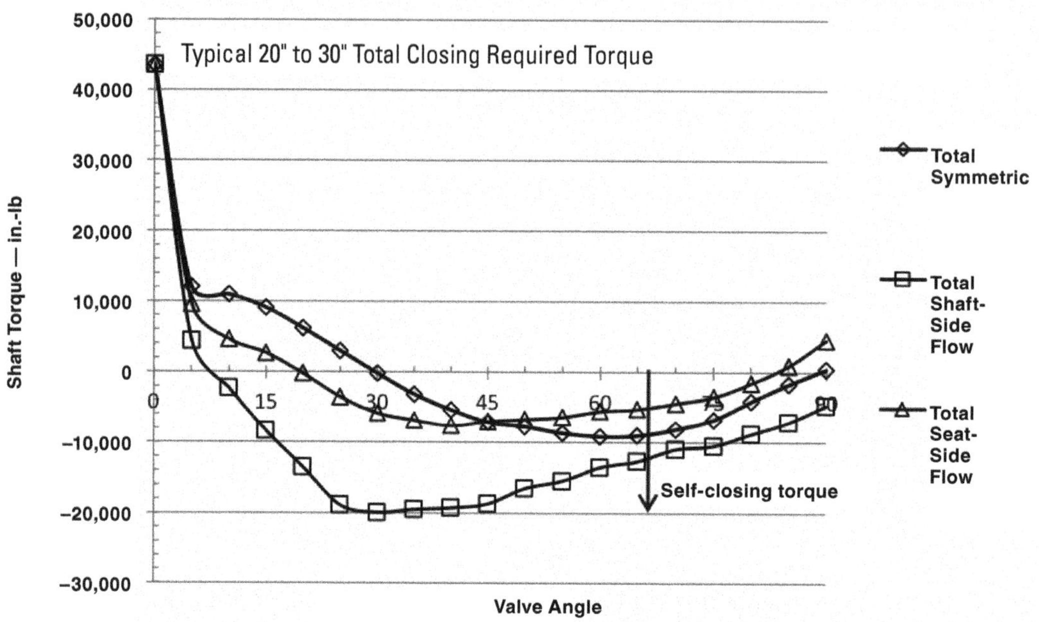

Figure 3-20 Typical total closing torque (T_{tc}) for a 20-in. (500-mm) to 30-in. (750-mm) BFV with symmetric and offset discs (Note: Metric torque is 1,130 N-m per division)

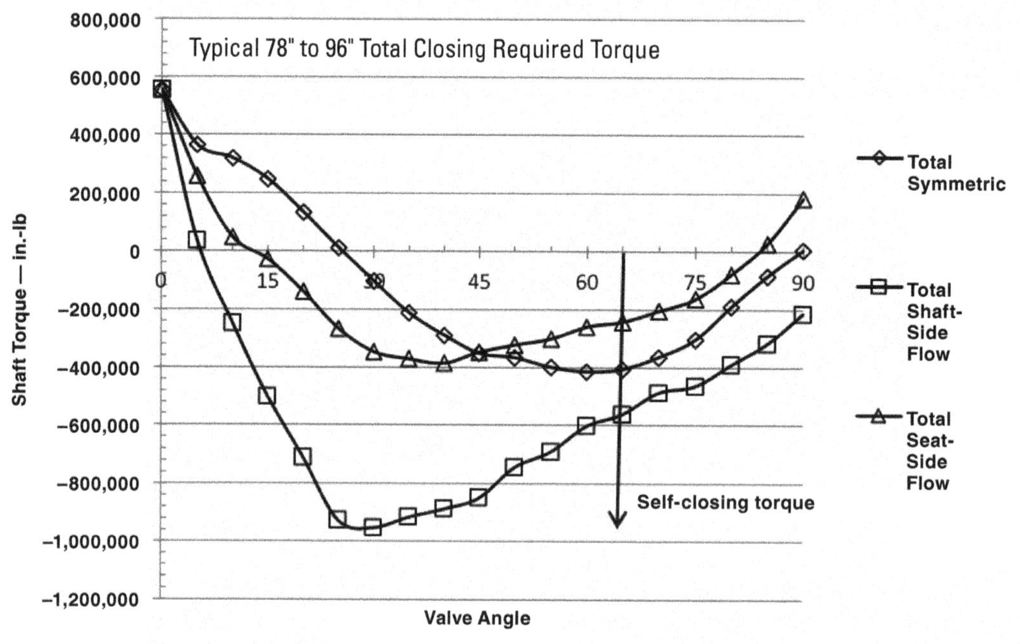

Figure 3-21 Total closing torque (T_{tc}) for a 78-in. (1,950-mm) to 96-in. (2,400-mm) BFV with symmetric and offset discs (Note: Metric torque is 22,600 N-m per division)

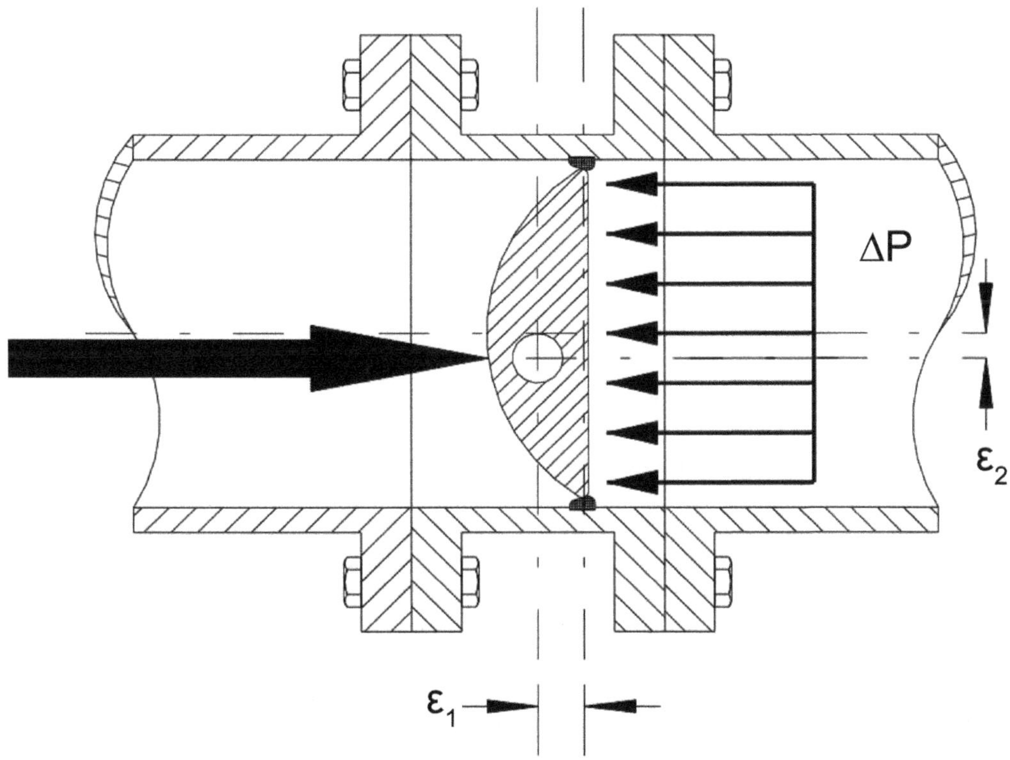

Figure 3-22 Shaft offset or eccentricity torque (Tecc)

(Ball, butterfly, or rotary cone); $$T_{ecc} = S_c \times U_{c2} \times \frac{\pi \times D_d^2 \times \varepsilon_2 \times \Delta P_\upsilon}{4}$$ (Eq 3-27)

or:

(Plug); $$T_{ecc} = S_c \times U_{c2} \times A_{port} \times \varepsilon_2 \times \Delta P_\upsilon$$ (Eq 3-28)

Where:

OTHER COMPONENTS OF TORQUE

The elements of torque described in the preceding sections apply to most quarter-turn valve applications. In certain designs, installations, and sizes, calculation of other torques or more exacting analysis may be needed. Detailed explanations of these methods are normally beyond the scope of this manual. Some components and methods are described here for clarification and convenience when needed. This type of analysis is not generally

Variable	Definition or Description	Units US Customary (SI—Metric)
A_{port}	Port area of PV	in. (mm)
D_d	Disc diameter, use nominal diameter (D) or D_{port} for BVs, RCVs, and PVs	in. (mm)
S_c	Sign convention variable: For torque active components, +1 when the torque tends to close the valve, or –1 when the torque tends to open the valve. For center-of-gravity torque, the sign convention variable is +1 when the center of gravity is above the horizontal when the closure member is in the full-open position or –1 when the center of gravity is below the horizontal when closure member is fully open. For torque transmitted to the actuator: a positive value when valve shaft torque opposes actuator motion or negative value when valve shaft torque assists actuator motion (actuator is acting as a brake).	dimensionless
T_{ecc}	Eccentricity torque (Positive value tends to close the valve; negative value tends to open valve.) NOTE: only considered at the seated position during opening or at closing	in.-lb (or ft-lb) (N-m)
U_{c2}	Units' conversion factor: US customary for torque in in.-lb: U_{c2} = 1 in./in. US customary for torque in ft-lb: U_{c2} = 1⁄12 (0.0833) ft/in. Metric for torque in N-m: $U_{c2} = 1 \times 10^{-6}$ (0.000001) m^2/mm^2	in./in. (ft/in.) (m^2/mm^2)
ΔP_v	Pressure drop (or loss) across the valve, general form	psid (kPa-d)
ε_2	Closure member lateral offset NOTE: ε_2 equals 0 for symmetric or single-offset closure member designs. Sign convention note: For hydrostatic torque, ε_2 is positive when oriented above the valve centerline and negative when oriented below the valve centerline. See Figure 3-3.	in. (mm)

possible during the design phase of a project, as the exact details and orientations are not known or available.

Bearing Torque Caused by Closure member and Shaft Assembly Weight, T_{bw}

The following is a more precise method for determining the bearing friction torque caused by the closure member and shaft weight. This is not normally used for simplicity but has been included for critical applications which often involve the consideration of large seismic accelerations. This result may be added to the bearing friction torque caused by differential pressure only (Eq 3-13 or 3-17).

The bearing torque from the closure member and shaft weight (T_{bw}) in a quarter-turn valve is a function of the coefficient of friction between the bearing and the shaft or trunnion, the shaft or trunnion diameter, the closure member and shaft (banjo) weight, and the orientation angle of pipe and shaft with respect to the vertical axis (see Figure 3-23).

$$T_{bw} = U_{c2} \times \frac{W_{d\&s} \times \sqrt{\text{COS}^2(\Omega) + (\text{SIN}(\Omega) \times \text{SIN}(\phi))^2} \times C_f \times d_s}{2} \quad \text{(Eq 3-29)}$$

Where:

Variable	Definition or Description	Units US Customary (SI—Metric)
A_{port}	Port area of PV	in.^{2} (mm)^2
C_f	Coefficient of friction between the shaft or trunnion and bushing, dimensionless (This value may be obtained from a flow test, engineering handbooks, the bearing manufacturer, or the valve manufacturer.)	dimensionless
D_d	Disc diameter, use nominal diameter (D) or D_{port} for BVs, RCVs, and PVs	in. (mm)
d_s	Shaft diameter	in. (mm)
T_{bw}	Bearing torque from closure member weight relative to installation orientation (always positive)	in.-lb (or ft-lb) (N-m)
$T_{b\theta}$	Bearing torque at valve angle θ (always positive)	in.-lb (or ft-lb) (N-m)
U_{c2}	Units' conversion factor: US customary for torque in in.-lb: U_{c2} = 1 in./in. US customary for torque in ft-lb: U_{c2} = 1⁄12 (0.0833) ft/in. Metric for torque in N-m: $U_{c2} = 1 \times 10^{-6}$ (0.000001) m^2/mm^2	in./in. (ft/in.) (m/mm)
$W_{d\&s}$	Weight of the closure member and shaft(s) assembly (banjo). For applications involving additional seismic loading, $W_{d\&s}$ may be multiplied by G or (G±1) where G is the additional gravitational acceleration multiplier.	lb (kg)
ΔP_θ	Pressure drop (or loss) while at closure member angle θ	psid (kPa-d)
ϕ	Valve installed shaft angle from vertical axis, 0–90 degrees (or 0 to $\pi/2$ rad)	degrees (rad)
Ω	Pipe angle from vertical axis for hydrostatic and bearing torque, 0–90 degrees (or 0 to $\pi/2$ rad) for flow downhill; 90–180 degrees (or $\pi/2$ to π rad) for flow uphill	degrees (rad)

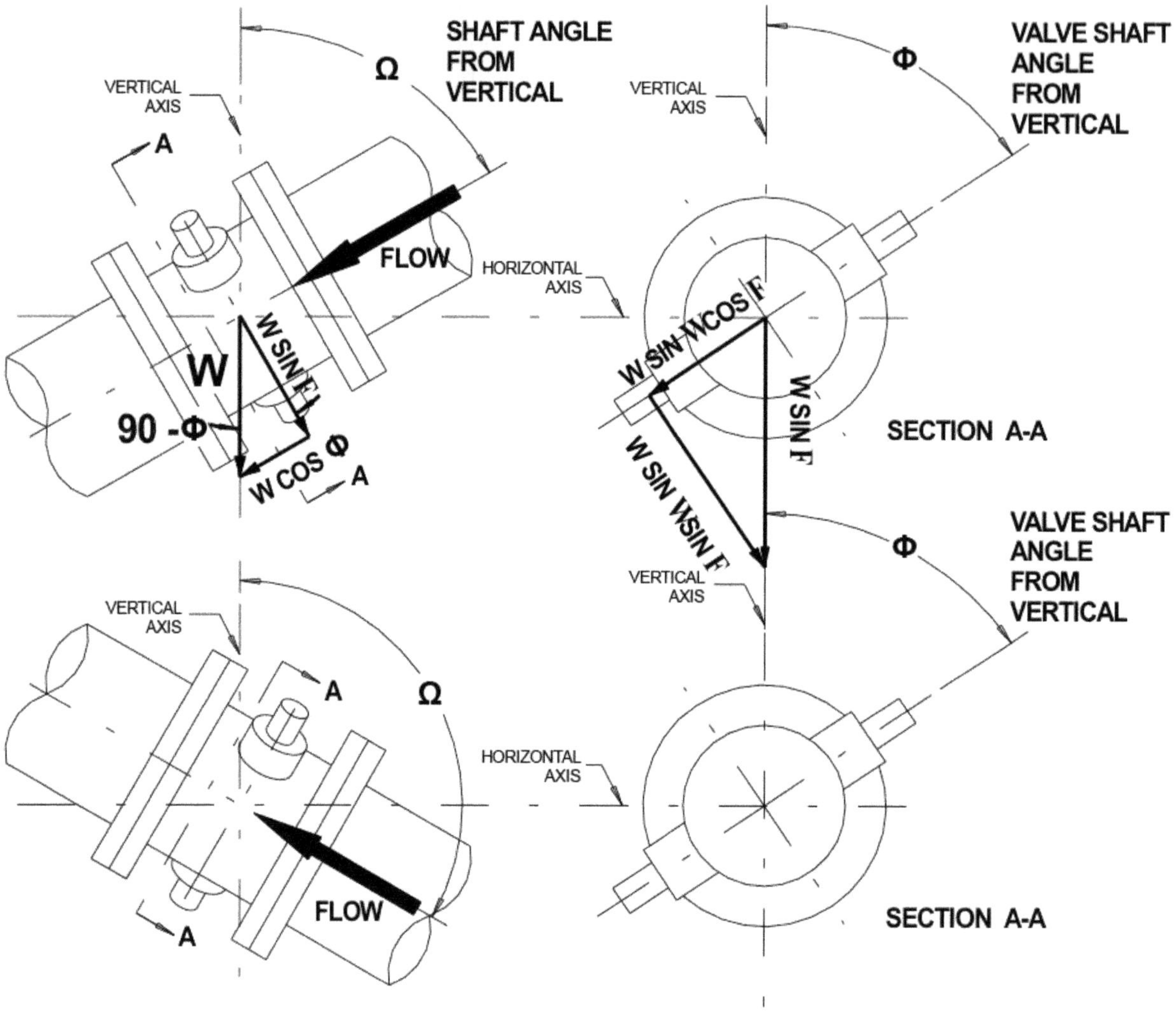

Figure 3-23 Bearing torque caused by closure member and shaft(s) weight orientation angles

The vector summation with the total bearing friction torque from both the loads caused by differential pressure and closure member and shaft(s) weight is as follows:

$$T_{btotal\theta} = (T_{b\theta} + T_{bw}) \quad \text{(Eq 3-30)}$$

Where:

Variable	Definition or Description	Units US Customary (SI—Metric)
$T_{btotal\theta}$	Total bearing torque at valve angle θ with addition of closure member and shaft(s) weight relative to installation orientation (always positive)	in.-lb (or ft-lb) (N-m)
T_{bw}	Bearing torque from closure member and shaft(s) weight relative to installation orientation (always positive)	in.-lb (or ft-lb) (N-m)
$T_{b\theta}$	Bearing torque at valve angle θ (always positive)	in.-lb (or ft-lb) (N-m)

Center-Of-Gravity Torque for Installed Orientation, $T_{CG\theta}$

For installations in which the pipe axis and shaft axis are known to be in other orientations, the following equations may be used. This more involved method is generally used for the more critical applications and often involves the consideration of large seismic accelerations.

$$T_{cg\theta} = U_{c1} \times S_c \times W_d \times C_g \times \text{SIN}(\beta - \theta - \gamma) \times \text{SIN}(\Phi) \qquad \text{(Eq 3-31)}$$

Where:

Variable	Definition or Description	Units US Customary (SI—Metric)
C_g	Valve closure member center-of-gravity distance from shaft centerline	in. (mm)
S_c	Sign convention variable: For torque active components, +1 when the torque tends to close the valve, or –1 when the torque tends to open the valve. For center-of-gravity torque, the sign convention variable is +1 when the center of gravity is above the horizontal when the closure member is in the full-open position or –1 when the center of gravity is below the horizontal when closure member is fully open. For torque transmitted to the actuator: a positive value when valve shaft torque opposes actuator motion or negative value when valve shaft torque assists actuator motion (Actuator is acting as a brake).	dimensionless
$T_{cg\theta}$	Center-of-gravity torque at valve angle θ (Positive value tends to close the valve; negative value tends to open the valve.)	in.-lb (or ft-lb) (N-m)
U_{c1}	Units' conversion factor: US customary for torque in in.-lb: $U_{c1} = 1$ in./in. US customary for torque in ft-lb: $U_{c1} = 1/12$ (0.0833) ft/in. Metric for torque in N-m: $U_{c1} = 1 \times 10^{-3}$ (0.001) m/mm	in./in. (ft/in.) (m/mm)
W_d	Weight of valve closure member. For applications involving additional seismic loading, $W_{d\&s}$ may be multiplied by G or (G±1) where G is the additional gravitational acceleration multiplier.	lb (kg)
β	Pipe angle from vertical axis for center of gravity relative to seat location, 0–90 degrees (or 0 to $\pi/2$ rad) when seated position is above horizontal; >90–180 degrees (or $\pi/2$ to π rad) when seated position is below horizontal	degrees (rad)
γ	Center-of-gravity offset angle (nonsymmetric closure member designs) May also include an adjustment for valve designs where the seating angle is not perpendicular to the pipe axis	degrees (rad)
θ	Valve opening position angle, closed = 0 degrees (0 rad); fully open 90 degrees ($\pi/2$ rad)	degrees (rad)
ϕ	Valve installed shaft angle from vertical axis, 0–90 degrees (or 0 to $\pi/2$ rad)	degrees (rad)

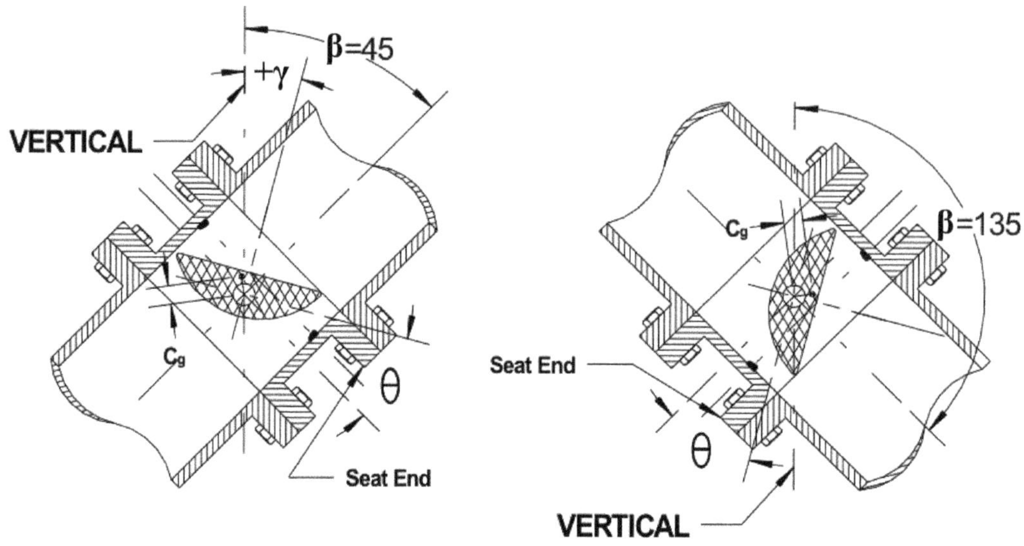

Figure 3-24 Center-of-gravity torque pipe angle definition

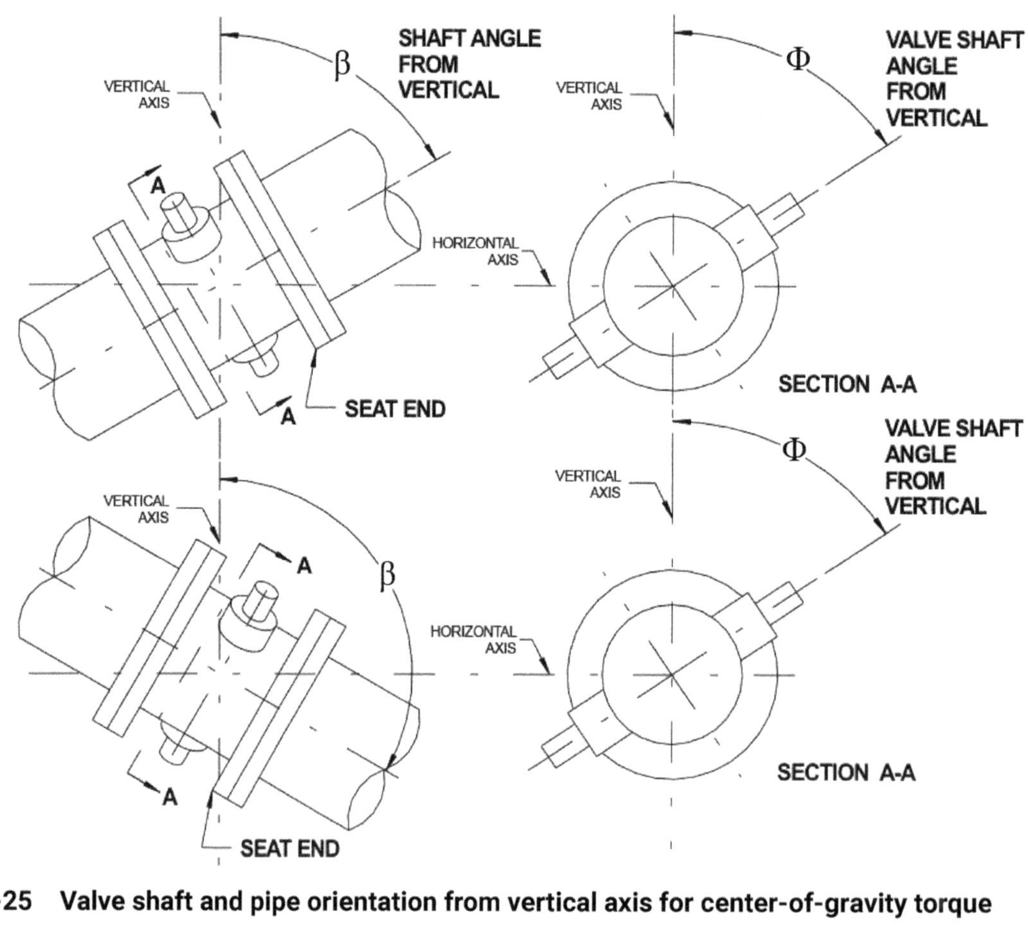

Figure 3-25 Valve shaft and pipe orientation from vertical axis for center-of-gravity torque

Hydrostatic Torque for Installed Orientation, Th

When the valve shaft is not vertical and the pipe axis is not horizontal but known, the following equations may be used. This more exact method is only necessary in critical analysis and requires a known installation orientation not generally available during the early design phases of a project.

(In US customary units);

$$T_h = S_c \times U_{c1} \times \frac{\rho \times \pi}{5.333} \times \left(\frac{D_d}{12}\right)^4 \times \left(\text{SIN}(\phi) + \frac{8 \times \varepsilon_2}{D_d}\right) \times \text{SIN}(\Omega) \qquad \text{(Eq 3-32)}$$

For cold water where ρ = 62.4 lb/ft^3:

(In US customary units);

$$T_h = S_c \times U_{c1} \times 0.0017726 \times (D_d)^4 \times \left(\text{SIN}(\phi) + \frac{8 \times \varepsilon_2}{D_d}\right) \times \text{SIN}(\Omega) \qquad \text{(Eq 3-33)}$$

For cold water where ρ = 1,000 kg/m^3:

(In SI metric units);

$$T_h = S_c \times 481.322 \times (U_{c1} \times D_d)^4 \times \left(\text{SIN}(\phi) + \frac{8 \times \varepsilon_2}{D_d}\right) \times \text{SIN}(\Omega) \qquad \text{(Eq 3-34)}$$

Where:

Variable	Definition or Description	Units US Customary (SI—Metric)
D_d	Disc diameter, use nominal diameter (D) or D_{port} for BVs, RCVs, and PVs	in. (mm)
g	Gravitational constant Acceleration due to gravity, 32.2 ft/s^2 (9.81 m/s^2)	ft/s^2 (m/s^2)
S_c	Sign convention variable: For torque active components, +1 when the torque tends to close the valve, or –1 when the torque tends to open the valve. For center-of-gravity torque, the sign convention variable is +1 when the center of gravity is above the horizontal when the closure member is in the full-open position and –1 when the center of gravity is below the horizontal when closure member is in the full-open position. For torque transmitted to the actuator: a positive value when valve shaft torque opposes actuator motion or negative value when valve shaft torque assists actuator motion (Actuator is acting as a brake.)	dimensionless
T_h	Hydrostatic torque (Positive value tends to close the valve; negative value tends to open the valve.) NOTE: only considered at the seated position during opening or at closing	in.-lb (or ft-lb) (N-m)

Variable	Definition or Description	Units US Customary (SI—Metric)
U_{c1}	Units' conversion factor: US customary for torque in in.-lb: U_{c1} = 1 in./in. US customary for torque in ft-lb: U_{c1} = 1⁄12 (0.0833) ft/in. Metric for torque in N-m: $U_{c1} = 1 \times 10^{-3}$ (0.001) m/mm	in./in. (ft/in.) (m/mm)
ε_2	Closure member lateral offset NOTE: ε_2 equals 0 for symmetric or single-offset closure member designs. Sign convention note: For hydrostatic torque, ε_2 is positive when oriented above the valve centerline and negative when oriented below the valve centerline. See Figure 3-3.	in. (mm)
ρ	Fluid density NOTE: Standard water density is considered as 62.4 lb/ft^3 (1,000 kg/m^3) at 40°F (4.4°C). Also see Table 4 in the Nomenclature section. Refer to Crane (2009 or later), Flow of Fluids Through Valves, Fittings and Pipe, Technical Paper No. 410	lb/ft^3 (kg/m^3)
ϕ	Valve installed shaft angle from vertical axis, 0–90 degrees (or 0 to $\pi/2$ rad)	degrees (rad)
Ω	Pipe angle from vertical axis for hydrostatic and bearing torque; 0–90 degrees (or 0 to $\pi/2$ rad) for flow downhill and 90–180 degrees (or $\pi/2$ to π rad) for flow uphill	degrees (rad)

NOTE: ε_2 is equal to zero (0) for symmetric and single-offset valves. The sign of ε_2 is positive when oriented above the valve centerline and negative when oriented below the valve centerline.

If the valve shaft angle, ϕ, equals zero (shaft vertical) or the pipe angle, Ω, equals zero (pipe vertical), then $T_h = 0$.

Buoyancy Torque

Another component of torque is generated by the buoyancy force of water displaced by the closure member acting vertically upward at its center of buoyancy. This torque is essentially opposite of the center-of-gravity torque and is generally low enough to be ignored except in very large (e.g., >60-in. [>1,500-mm]), low-pressure (e.g., <25-psig [<172-kPa]) designs with hollow closure member structures. This methodology is similar to the weight and center-of-gravity torque.

Thrust Bearing Torque

This component of torque is generated by frictional resistance of the valve thrust bearing, which centers the closure member axially along the shaft while supporting the closure member and shaft weight, and the shaft thrust due to internal pressure (i.e., piston effect or shaft ejection thrust). This torque is generally negligible except in very high-pressure designs (e.g., 250-psig [1.72-MPa]) and high-friction thrust bearing designs.

Upstream Flow Disturbances Flow and Asymmetric Flow Distributions

Unusual installations in which a quarter-turn valve is mounted downstream of elbows, pumps, or other valves, for example, may cause unusually high or unstable dynamic

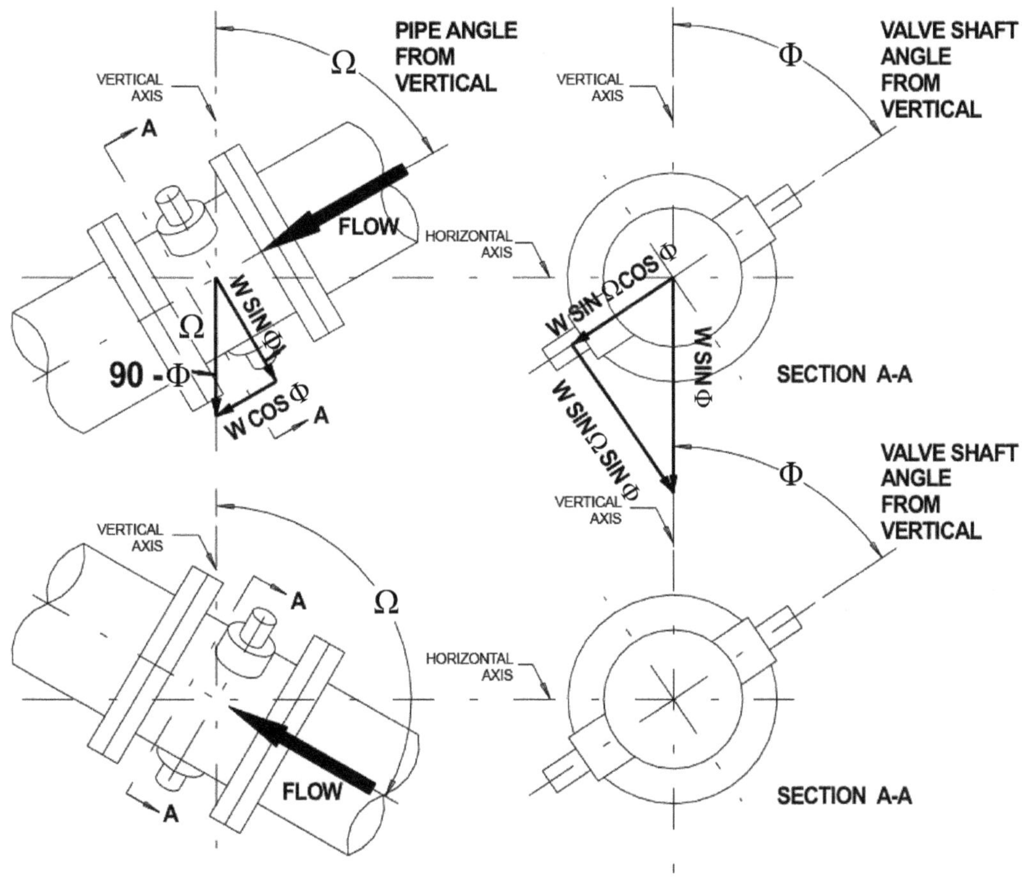

Figure 3-26 Valve shaft and pipe orientation from vertical axis for hydrostatic and bearing torque

torques if the valves are not oriented properly. See additional explanations in Chapter 6. If a special orientation is required, flow tests of the actual piping configuration, or a model of it, can be conducted to determine an applicable set of flow and torque coefficients. Also, some researchers have developed approximate methods for estimating the amount of dynamic torque increase in relation to upstream elbows at various installed orientations.

SYSTEM CHARACTERISTICS

As discussed in Chapter 2, system characteristics must be known to calculate the flow, ΔP, and torque for each valve position as the valve is opened or closed. This is done using the equivalent resistance system model (ERSM). Pressure drop in a piping system is caused by friction losses in the pipe, valves, and other components of the system. The system flow rate increases or decreases with changes in the valve position. As the flow rate increases or decreases, the friction losses in the remainder of the system change in relation to the square of the change in flow rate. For example, reducing the flow rate by one-half causes the friction loss in the pipe to decrease to one-fourth of the original value. Given this relationship, the system design data and the initial flow parameters must be known to determine the pressure drop across the valve and the torque values as the valve position changes.

The information needed for a complete model and analysis is as follows.

Description of the System Head Source (Constant or Variable)

If the system head is variable, the pump curve should be included. If the pump curve is not available, a constant-head source with a source pressure equal to the closed differential pressure across the valve will be assumed.

Maximum Pressure Differential

Maximum differential pressure across the closed valve

Maximum Design Flow Rate

Maximum design flow through the piping system through the fully open valve

System Pressure Drop

System pressure drop at the design flow rate. If this information is not provided, a pressure drop equal to the closed pressure differential across the valve is assumed.

System Analysis and Example Calculations

Figure 2-6 shows a basic diagram of important relationships. Calculations involve the following variables:

Variable	Definition or Description	Units US Customary (SI—Metric)
ΔH	Head loss between any two points in a system	feet of water (meters of water)
ΔP	Pressure drop (or loss) between any two points in a system	psi (kPa)
K	Resistance coefficient	dimensionless
K_{v90}	Resistance coefficient of valve at full open (90 degrees)	dimensionless
g	Gravitational constant (acceleration due to gravity)	ft/s^2 (m/s^2)
D	Valve diameter	in. (m)
V	Velocity of flow	feet per second, fps, or ft/s (meters per second, mps, or m/s)

The additional variable subscripts that are applied to variables such as C_v, T_d, T_b, ΔP, ΔH, or K are:

Variable	Definition or Description
sys	Subscript indicating system piping and components less the BFV
V	Subscript indicating the quarter-turn valve
0 through 90	Subscript indicating the valve closure member angle

These methods are based on an equivalent resistance model as graphically shown in Figure 2-6.

Torque Calculation Methodology Example

Given the system data and calculations from Chapter 2 and the additional valve data as follows:

- Valve input data
  - 24-in. (600-mm) AWWA Class 150B BFV with 24-in. (600-mm) diameter single-offset disc, seat-side flow; valve is installed in a horizontal line with a vertically oriented shaft.
  - $K_{v\theta}$ from Chapter 2, Table 2-1
  - $C_{t\theta}$ from Table 3-2
  - C_f for a bronze bearing = 0.25
  - d_s = 3.00 in. (76.2 mm) (per ANSI/AWWA C504)
  - T_p = 1,350 in.-lb (152 N-m)
  - D_d = 24.0 in. (609 mm)
  - T_{cg} = 0 in.-lb (0 N-m)
  - $C_{sc} = C_{usc}$ = 16.0 lb/in. (2,802 mm)
  - $C_{sp} = C_{usp}$ = 0.03 lb/in./psi (0.762 mm)
  - $W_{d\&s}$ = 450 lb (204 kg)
  - On-off electric motor AF from AWWA C504 = 1.25

1. From the inputs, T_{cg} and T_h are not applicable because the shaft is vertically oriented. T_{ecc} is not applicable because the design is a single-offset BFV.
2. From Chapter 2, calculate the valve differential pressures and flow rates at the valve angle (θ).
3. Calculate dynamic torque $T_{d\theta}$ at the valve angle (θ):

$$T_{d\theta} = U_{c2} \times C_{t\theta} \times D_d^3 \times DP_{dyn\theta} \qquad \text{(Eq 3-26)}$$

4. Calculate bearing torque $T_{b\theta}$ at the valve angle (θ):

$$T_{b\theta} = U_{c2} \times \frac{\left(\pi \times D_d^2 \times DP_{brg\theta} + W_{d\&s}\right) \times d_s \times C_f}{8} \qquad \text{(Eq 3-14)}$$

NOTE: This equation is for bearing torque due to differential pressure plus disc and shaft weight as a direct summation.

5. Calculate the seating and unseating torque, T_s and T_{us}, at the valve angle (0):

$$\text{(Ball, butterfly);} \qquad T_s = U_{c2} \times \left(C_{sc} + C_{sp} \times \Delta P_{max}\right) \times D_d^2 \qquad \text{(Eq 3-5)}$$

6. Calculate total torque $T_{t\theta}$ at the valve angle for the opening and closing directions

$$\text{(At valve angle 0);} \qquad T_{ts} = T_{b0^o} - T_{cg0^o} - T_h + T_s + T_p - T_{ecc}, \qquad \text{(Eq 3-1)}$$

$$\text{(At valve angle 0);} \qquad T_{tus} = T_{b0^o} + T_{cg0^o} + T_h + T_{us} + T_p + T_{ecc}, \qquad \text{(Eq 3-2)}$$

$$\text{(At valve angles >0);} \qquad T_{to\theta} = T_{b\theta} + T_{cg\theta} + T_{d\theta} + T_p, \qquad \text{(Eq 3-3)}$$

$$\text{(At valve angles >0);} \qquad T_{tc\theta} = T_{b\theta} - T_{cg\theta} - T_{d\theta} + T_p, \qquad \text{(Eq 3-4)}$$

Note that the T_d torque will be added in the opening direction and subtracted in the closing direction.

7. Repeat steps 2 through 6 for other valve angles.

Figure 3-27 shows a graphical representation of the calculations in this example.

Table 3-2 Torque calculation data for constant-head source example

Input	Ch. 2 Calc.		Input	Calc.	Calc.	Calc.	Calc.	Calc.	Calc.	Calc.
Valve Angle θ *deg.*	Valve Loss $\Delta H_{v\theta}$ *ft* *(m)*	Valve Loss $\Delta P_{v\theta}$ psid *(kPa-d)*	Dynamic Torque Coef. $C_{t\theta}$	Dynamic Torque $T_{d\theta}$ *in.-lb* *(N-m)*	Bearing Friction Torque $T_{b\theta}$ *in.-lb* *(N-m)*	Seating Torque T_s *in.-lb* *(N-m)*	Packing Friction Torque $T_{p\theta}$ *in.-lb* *(N-m)*	Total Opening Torque $T_{to\theta}$ [MRST] *in.-lb* *(N-m)*	Total Closing Torque $T_{tc\theta}$ [MRST] *in.-lb* *(N-m)*	AST *in.-lb* *(N-m)*
90	0.9	0.41	–0.3210	–1,809	111	0	1,350	–347	3,270	4,087
	(0.27)	2.83		(–204)	(12.5)	(0)	(153)	(–39.2)	(369)	(462)
80	1.2	0.54	0.0969	726	134	0	1,350	2,210	758	2,762
	(0.37)	3.72		(82.0)	(15.1)	(0)	(152.5)	(249.7)	(85.6)	(312.1)
70	3.4	1.46	0.1250	2,519	290	0	1,350	4,159	–880	5,198
	(1.04)	10.1		(284.6)	(32.8)	(0)	(152.5)	(469.9)	99.4	(587.3)
60	8.9	3.87	0.0800	4,282	699	0	1,350	6,331	–2,233	7,914
	(2.71)	26.7		(483.8)	(79.0)	(0)	(152.5)	(715.3)	252.3	(894.2)
50	20.8	9.02	0.0511	6,369	1,572	0	1,350	9,290	–3,447	11,613
	(6.34)	62.2		(719.6)	(177.6)	(0)	(152.5)	(1,049.6)	389.5	(1,312.1)
40	44.0	19.06	0.0341	8,984	3,275	0	1,350	13,610	–4,359	17,012
	(13.41)	131.4		(1,015.1)	(370.0)	(0)	(152.5)	(1,537.7)	492.5	(1,922.1)
30	72.5	31.43	0.0219	9,514	5,373	0	1,350	16,237	–2,791	20,297
	(22.10)	216.7		(1,074.9)	(607.1)	(0)	(152.5)	(1,834.5)	315.3	(2,293.3)
20	91.3	39.60	0.0128	7,006	6,759	0	1,350	15,116	1,103	18,894
	(27.83)	273.0		(791.6)	(763.7)	(0)	(152.5)	(1,707.9)	(124.6)	(2,134.7)
10	99.0	42.90	0.0080	4,744	7,320	0	1,350	13,414	3,925	16,767
	(30.18)	295.8		(536.0)	(827.1)	(0)	(152.5)	(1,515.6)	(443.5)	(1,894.4)
0	100.0	43.35	0.0000	0	7,396	9,965	1,350	18,711	18,711	23,389
	(30.48)	298.9		(0)	(835.6)	(1,125.9)	(152.5)	(2,114.1)	(2,114.1)	(2,642.6)

Calc.: calculation, Ch.: Chapter, Coef.: coefficient

NOTES:

The data are for the example only and have no relationship to an actual valve.

Positive $C_{t\theta}$ and $T_{d\theta}$ values indicate a dynamic torque that is acting to close the valve. Negative $C_{t\theta}$ and $T_{d\theta}$ values indicate a dynamic torque that is acting to open the valve.

Negative values for valve closing torque indicate that the valve is self-acting to close at those positions; thus, the valve actuator must resist this torque as a brake.

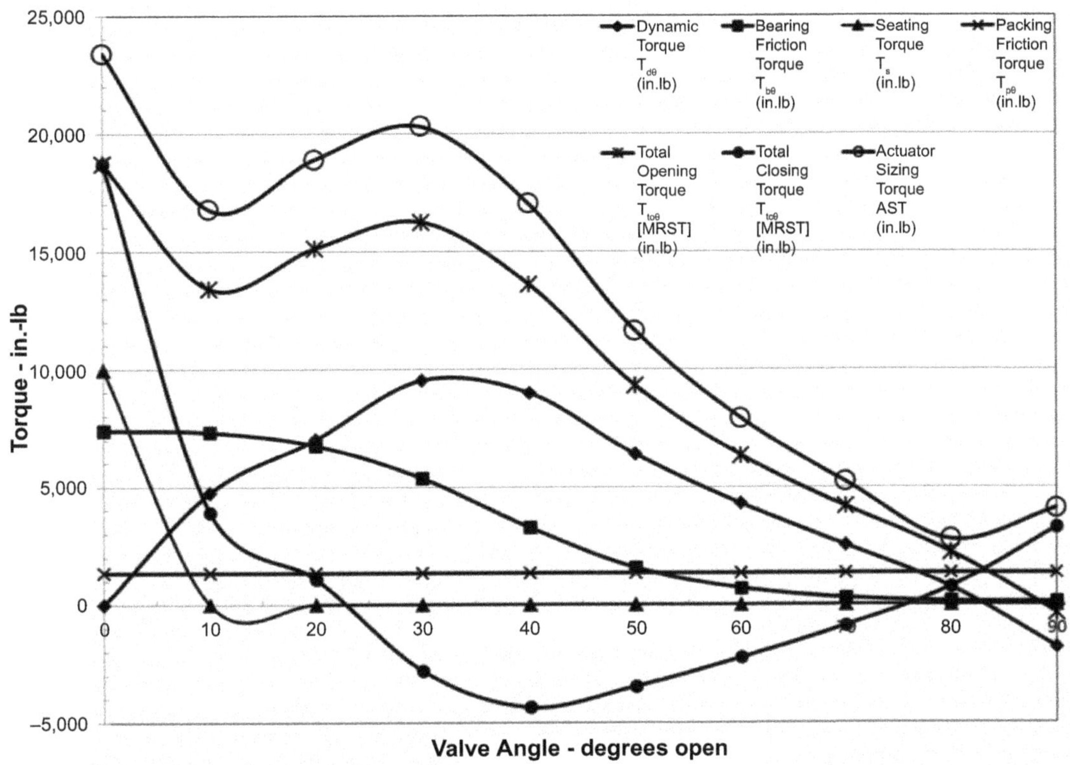

Figure 3-27 Example torque calculation summary graph (NOTE: Metric torque is 565 N-m per division)

REFERENCES

ANSI/AWWA. 2023. C504. Rubber-Seated Butterfly Valves. Denver: AWWA.

ANSI/AWWA. 2023. C507. Ball Valves, 6 in. Through 60 in. (150 mm through 1,500 mm). Denver: AWWA.

ANSI (American National Standards Institute)/AWWA. 2021. C516. Large Diameter Rubber-Seated Butterfly Valves, 78 in. (2,000 mm) and Larger. Denver: AWWA.

ANSI/AWWA. 2021. C517. Resilient-Seated Cast-Iron Eccentric Plug Valves. Denver: AWWA.

ANSI/AWWA. 2021. C541. Hydraulic and Pneumatic Cylinder and Vane-Type Actuators for Valves and Slide Gates. Denver: AWWA.

ANSI/AWWA. 2021. C542. Electric Motor Actuators for Valves and Slide Gates. Denver: AWWA.

Crane. 2018. "Flow of Fluids Through Valves, Fittings and Pipe, Technical Paper No. 410." Crane.

This page intentionally blank.

Chapter **4**

Valve Cavitation and Choking

When a quarter-turn valve is used for throttling or modulating flow rates, the operating conditions should be evaluated to determine whether significant cavitation will occur. Cavitation can cause objectionable noise and vibration and decrease the useful life of a valve and nearby piping components.

The topics in this chapter include an explanation of the conditions that cause cavitation, a method for predicting it, and a listing of methods for minimizing its effects. The choking calculations are also examined.

DEFINITIONS

Cavitation

Cavitation is the vaporization and subsequent violent condensation of a fluid caused by localized areas of low pressure in a piping system. When water flows through a partially open quarter-turn valve, a localized low-pressure zone may occur immediately downstream of the valve closure member because of the sudden changes in flow velocity and flow separation. When the pressure in this zone falls below the vapor pressure of the fluid, the liquid vaporizes, forming a vapor pocket or vapor bubbles. As the bubbles flow downstream and the pipeline pressure recovers, the bubbles violently collapse or implode. Bubble collapse near a boundary, valve component, fitting, or pipe wall may result in pitting and material removal. Measurements have shown that localized pressures of 100,000 psi (689 MPa) can be generated by the implosion of the bubbles. These rapid implosions can produce effects varying from a popping sound to rumbling or even a deafening roar approaching 100 dB (Tullis 1989). Finally, when cavitation is fully developed, flow is restricted and no longer proportional to the square root of differential pressure.

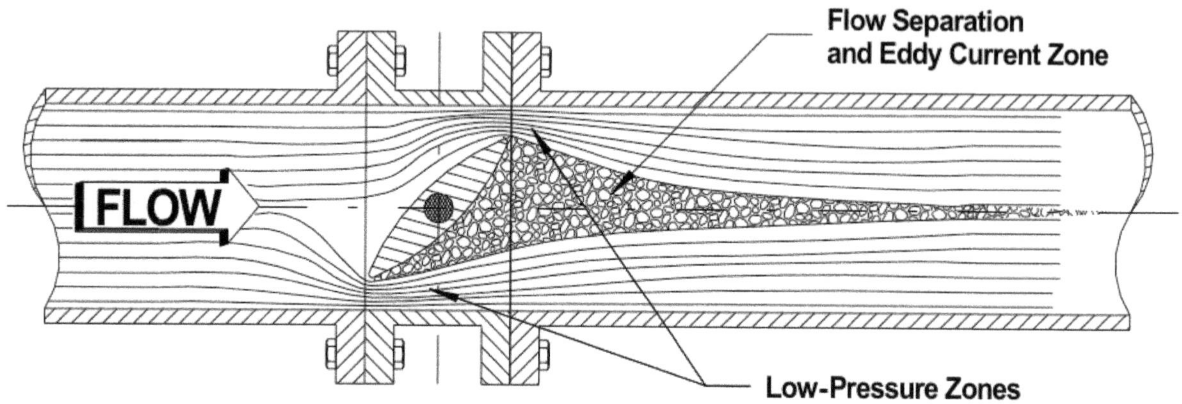

Figure 4-1 Cavitation zones downstream of a BFV disc

Cavitation can form immediately downstream of the quarter-turn valve closure member where a low-pressure zone occurs. Figure 4-1 illustrates the low-pressure areas downstream of a butterfly valve (BFV) disc. Cavitation bubbles can initiate near the disc edges, downstream of the body seat, in the downstream shadow of the disc, or near the downstream pipe wall. Cavitation bubbles can implode just downstream of the valve or many times the pipe diameter downstream, depending on where the pressure recovers. The process produces an unmistakable noise and vibration that sounds like gravel flowing through the pipe.

Many simple shutoff valve applications produce cavitation when the valves are near the closed position because the differential pressure reaches its highest level at that point. However, because a shutoff valve is usually at a near-closed angle for only a short period of time, appreciable damage to the valve or piping usually does not occur.

When a valve is exposed to cavitating conditions continuously, however, when it is used for flow modulation or pressure control, significant damage can occur to the metal surfaces of the valve or downstream piping in a short period of time. Hence, modulating and throttling applications warrant evaluation of cavitation conditions.

Three terms are commonly used to classify cavitation in valves according to the International Society of Automation (ISA-RP75.23, Considerations for Evaluating Control Valve Cavitation):

1. Incipient cavitation
2. Constant cavitation
3. Choked cavitation

The start of steady cavitation, termed incipient cavitation, can be indicated by an intermittent popping sound in the flow stream (Point A in Figures 4-2 and 4-3). Incipient cavitation typically does not cause damage or objectionably loud noise. If the pressure differential increases, however, the constant cavitation level is reached, which can be indicated by a continuous popping similar to the sound of gravel flowing through the pipe or bacon frying (Point B in Figure 4-3). Flowing conditions up to the constant cavitation conditions will not cause advanced damage or degradation to the valve or piping. Continuous flow above the constant cavitation level is often accompanied by objectionable noise and valve or piping damage. Finally, the choking cavitation level (Point C in Figures 4-2 and 4-3) occurs when the valve is passing the maximum flow possible for a given upstream pressure. The vapor pocket may become extremely long, causing damage far downstream from the valve. Choking cavitation may cause a reduction in noise, but this change is usually preceded by the highest level of noise and vibration. Valves operating at the choking

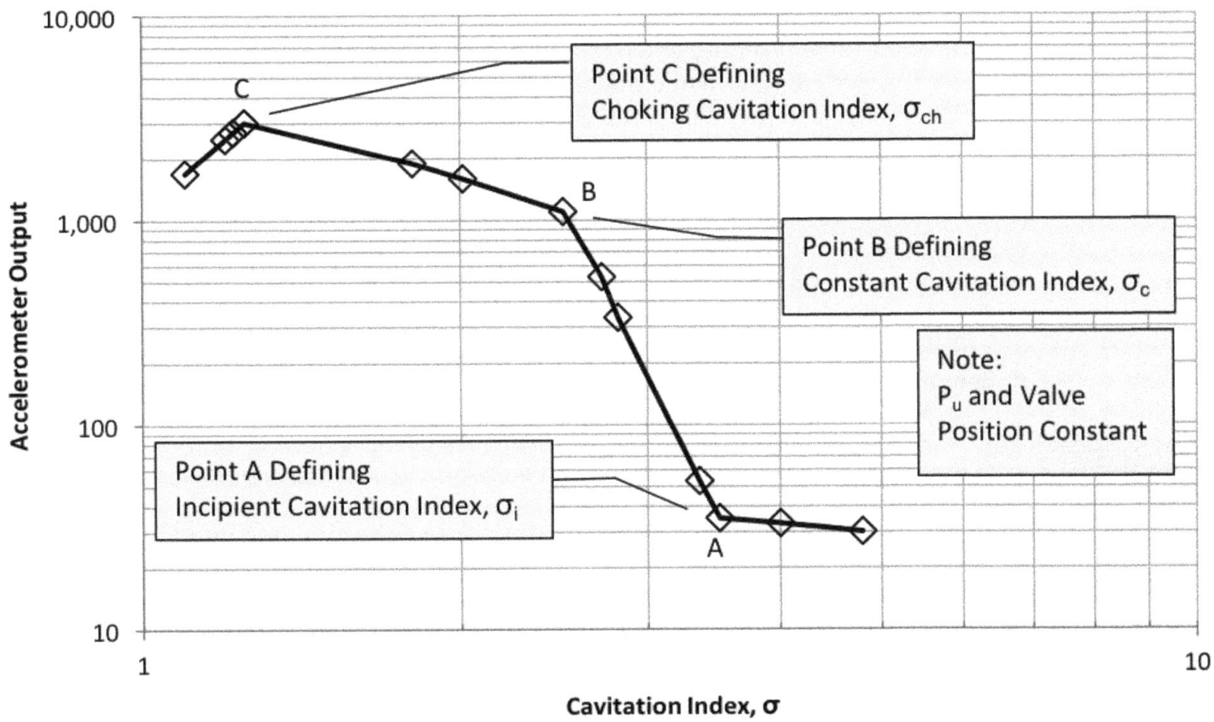

Figure 4-2 Typical cavitation index levels and acceleration readings (Tullis 1989, page 135)

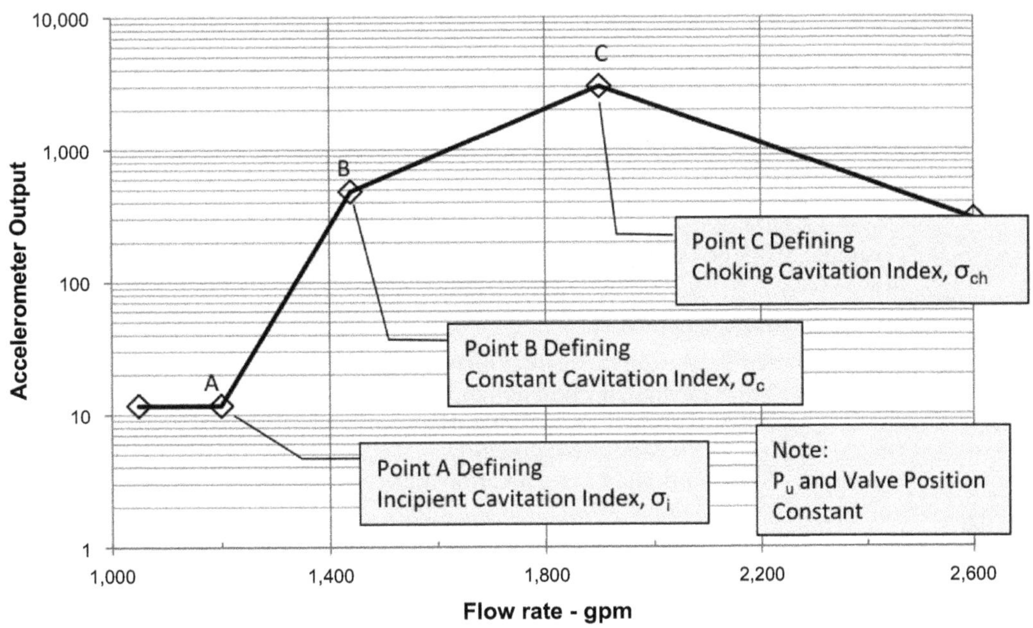

Figure 4-3 Flow rate and acceleration readings

cavitation level usually allow short bursts of flow accompanied by high velocities and potentially high operating torques. Conditions that produce choking should be reviewed with the valve manufacturer. Reference Figure 2-1 for an overall picture of when cavitation and choking can occur in an operating system.

Choked flow for incompressible fluids

At a constant valve position, when the differential pressure ΔP is increased by decreasing the downstream pressure while maintaining a constant upstream pressure, the flow rate, Q, through the valve increases based on the relationship between Q and $\Delta P_V^{1/2}$. This is shown as the linear trend in Figure 2-1 in the "Cavitation Free Region." The slope of the line defines the C_v for the valve at that position. As the downstream pressure is further reduced, the change in flow rate is no longer linear with respect to $\Delta P_V^{1/2}$ and eventually reaches a maximum value as shown near the F_L definition point on Figure 2-1. The flow is choked because of flash vaporization of the liquid when the pressure downstream of the closure member falls below the vapor pressure, P_{VAPOR}, of the liquid.

To simplify calculations, this is generally treated as two linear section meetings at the F_L definition point of Figure 2-1.

PREDICTING CAVITATION

Tests have shown that conditions likely to produce cavitation in a quarter-turn valve can be predicted and possibly reduced or prevented. The cavitation index is typically used as a predictor of valve damage and is expressed quantitatively at each valve angle as follows (ISA-RP75.23):

$$\sigma = \frac{P_u - P_v}{P_u - P_d} \quad \text{(Eq 4-1)}$$

Where:

Variable	Definition or Description	Units US Customary (SI—Metric)
P_d	Reference downstream pressure for cavitation analysis	psi (kPa)
P_u	Reference upstream pressure for cavitation analysis	psi (kPa)
P_v	Vapor pressure adjusted for temperature and atmospheric pressure (Example: P_v = –14.4 psig [–99.6 kPa] for water at 60°F [16°C], measured at sea level)	psi (kPa)
σ	Cavitation index, general form	dimensionless
σ subscripts	Subscripts used for the σ index values include: i Valve's incipient cavitation index c Valve's constant cavitation index ch Valves choking cavitation index Operating System installed and operating index	N/A

N/A: not applicable

NOTE: Pressures may be gauge or absolute but must be consistent. Pressure may also be taken as feet of head as long as consistent.

The operating cavitation index can be compared to the cavitation indexes for valves to predict what level of cavitation will occur (incipient, constant, or choking). In some earlier texts, the constant index is referred to as "critical." Later texts changed this nomenclature to "constant" to be more descriptive of the condition without implying an undesirable operating condition.

A valve's cavitation indexes for incipient, constant, and choked levels can be determined from flow testing in a laboratory environment. Cavitation can be observed

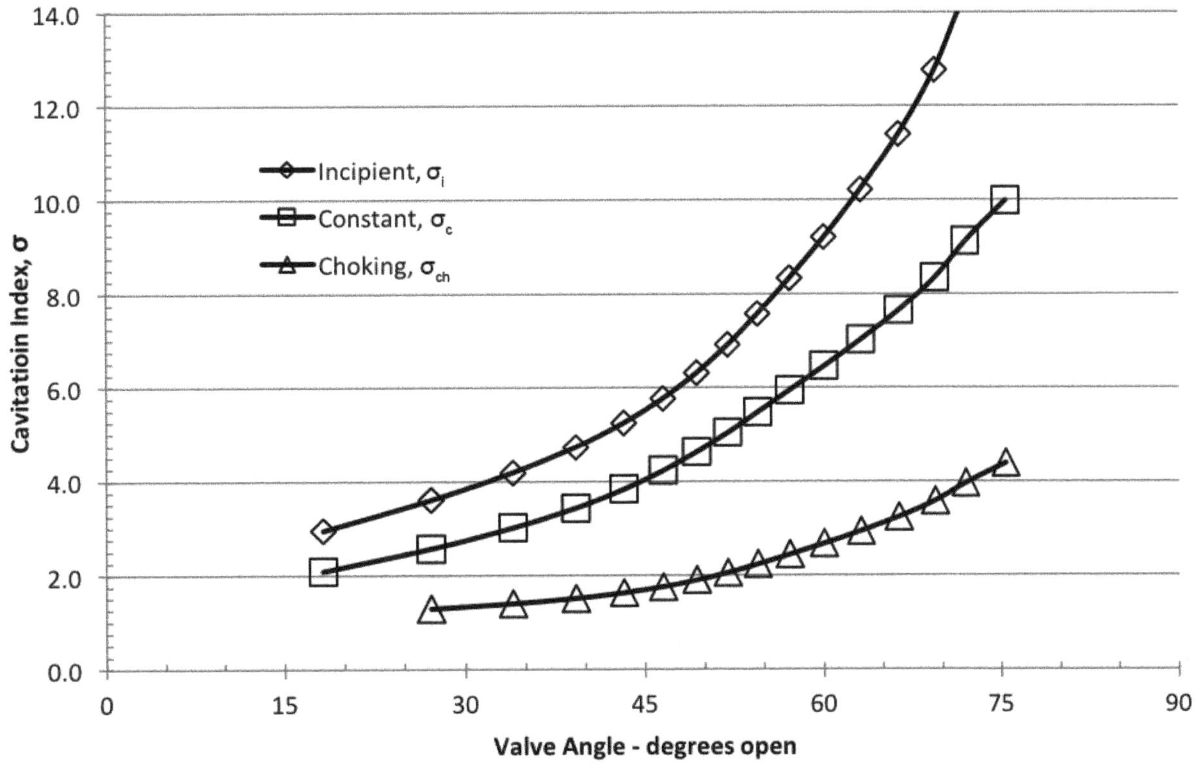

Figure 4-4 Typical cavitation index values for a 6-in. (150-mm) BFV (Reference upstream pressure from laboratory test = P_{ut} = 70 psi [483 kPa], vapor pressure from laboratory test = P_{vt} = −12 psi [−82.7 kPa]). Figure is based on Tullis (1989, page 140).

* "Critical" annotation was changed to "Constant" in the figure to be more descriptive of the condition and match current practice.

† The data were adjusted to upstream pressure correlation instead of downstream pressure reference to match current practice and the methodology used in this manual.

‡ The abscissa (x-axis) was modified from the flow coefficient, C_d, to the corresponding valve position angle for ease of use in this manual of practice.

using a hydrophone or accelerometer during the flow test. Audible detection by one with a trained ear can readily identify the incipient or constant levels of cavitation. The results of a flow test on a 6-in. (150-mm) BFV at a reference test pressure of 70 psi (483 kPa) are shown in Figure 4-4. Similar data can be prepared for any quarter-turn valve using the test methodology given in Chapter 5.

Once the valve is installed into a system and operating under a set of conditions, the operating or system index ($\sigma_{operating}$) can be calculated. The lower the calculated operating cavitation index (see Eq 4-1) value is, the greater the likelihood of cavitation damage is.

For example, if a valve is throttled at 45 degrees open with a calculated system or operating index of ($\sigma_{operating}$) 6.0, then cavitation will not likely occur (referencing Figure 4-4). If, however, the valve is closed further to 30 degrees open with a calculated system index ($\sigma_{operating}$) of 2.2, then the cavitation in the range between incipient and constant will occur. Sounds of cavitation will be heard, but serious damage will occur only after a prolonged period of time under those conditions.

Cavitation data are typically reported for a given valve size and upstream test pressure. Scale and pressure factors can be applied to the data to adjust coefficients from one size and pressure to another.

Incipient and constant cavitation indexes (σ_i and σ_c) can be corrected for size scale effects (SSEs) and pressure scale effects (PSEs) by these equations (Tullis 1989, pages 144–148; Tullis 1993, pages 47–55):

$$\sigma_i = (\sigma_{it} - 1) \times \text{PSE} \times \text{SSE} + 1 \quad \text{(Eq 4-2)}$$

$$\sigma_c = (\sigma_{ct} - 1) \times \text{PSE} \times \text{SSE} + 1 \quad \text{(Eq 4-3)}$$

$$\text{PSE} = \left(\frac{P_u - P_v}{P_{ut} - P_{vt}} \right)^{0.28}, \text{Note *} \quad \text{(Eq 4-4)}$$

$$\text{SSE} = \left(\frac{D}{d_t} \right)^{Y}, \text{Note}^{\dagger} \quad \text{(Eq 4-5)}$$

$$Y = 0.3 \times K_v^{-0.25} \quad \text{(Eq 4-6)}$$

NOTES:

* Results may be quite conservative for pressures above 300 psia.

† For SSE calculations, D is generally limited to ≤36 in. For valves greater than 36 in., it is believed that SSEs will be over predicted.

Where:

Variable	Definition or Description	Units US Customary (SI—Metric)
D	Nominal valve diameter	in. (mm)
d_t	Size of model or test valve	in. (mm)
K_v	Flow resistance coefficient of the valve	dimensionless
PSE	Pressure scale effects factor for cavitation analysis	dimensionless
P_u	Reference upstream pressure for cavitation analysis	psi (kPa)
P_{ut}	Upstream pressure from laboratory test for cavitation analysis	psi (kPa)
P_v	Vapor pressure adjusted for temperature and atmospheric pressure (Example: P_v = –14.4 psig [–99.6 kPa] for water at 60°F [16°C], measured at sea level)	psi (kPa)
P_{vt}	Vapor pressure from laboratory test	psi (kPa)
SSE	Sizing scale effects factor for cavitation analysis	dimensionless
Y	Size scale exponent for cavitation analysis	dimensionless
σ_c	Constant cavitation index at a reference pressure, P_u	dimensionless
σ_{ct}	Constant cavitation index from laboratory testing	dimensionless
σ_i	Incipient cavitation index at a reference pressure, P_u	dimensionless
σ_{it}	Incipient cavitation index from laboratory test	dimensionless
$\sigma_{operating}$	System installed and operating cavitation index	dimensionless

CAVITATION CALCULATION METHODOLOGY EXAMPLE

Given the system data and calculations from Chapter 2, the valve data from Chapter 3, and cavitation index values from Figure 4-4:

1. From the inputs, operating vapor pressure is –14.4 psig.
2. From Chapter 2, calculate the valve upstream and differential pressures at the valve angle (θ).
3. Calculate the valve's operating cavitation index, σ_θ, at valve angle (θ) using Eq 4-1.
4. Calculate the PSE factor, PSE_θ, at valve angle (θ) using Eq 4-4.
5. Calculate the size scale exponent, Y_θ, at valve angle (θ) using Eq 4-6.
6. Calculate the SSE factor, SSE_θ, at valve angle (θ) using Eq 4-5.
7. Calculate the reference-corrected incipient cavitation index, σ_i, at valve angle (θ) using Eq 4-2.
8. Calculate the reference-corrected constant cavitation index, σ_c, at valve angle (θ) using Eq 4-3.
9. Repeat steps 3 through 8 for other valve angles (θ).
10. Compare the valve's operating cavitation index to the reference-corrected incipient cavitation index and the reference-corrected constant cavitation index at each valve angle (θ). When the operating cavitation index is at or below the reference-corrected incipient cavitation index, cavitation starts and increases as the operating cavitation index drops further. When the operating cavitation index is between the reference-corrected incipient cavitation index and the reference-corrected constant cavitation index, cavitation exists and is audible within the system, but cavitation-related damage is unlikely. When the operating cavitation index is below the reference-corrected constant cavitation index, cavitation is constant and fully developed. As the operating cavitation index drops further, the risk of cavitation-related damage increases.

    Table 4-1 shows the calculated data, and Figure 4-5 shows a graphical representation for this example.

METHODS OF REDUCING CAVITATION

Design provisions for water systems completely without cavitation are beyond the scope of this manual, but some general recommendations to reduce cavitation can be considered. A detailed look at the cavitation index equation (Eq 4-1) may offer clues about how cavitation can be reduced.

To reduce cavitation, the value of the operating cavitation index, $\sigma_{operating}$, must be increased above the constant cavitation index for the valve, σ_c, similar to that shown in Figure 4-4. One way to do this is to increase the downstream pressure, P_d, which will increase the value of the cavitation index. Another strategy is to decrease the differential pressure across the valve ($P_u - P_d$). The value of the constant cavitation index, σ_c, can also be changed by using the valve at a different opening position or using a different valve type or model. This includes valves that incorporate cavitation-reducing trim that are available for many valves. Finally, air can be introduced to mitigate cavitation.

In practice, these changes can be achieved in some cases by using one or more of the following methods (Tullis 1989, pages 145–165; Skousen 1998, pages 511–517).

1. Increase the pressure downstream of the valve. This involves relocating the quarter-turn valve in the system or providing additional flow restriction downstream using another valve or permanent restriction such as an orifice.
2. Decrease the differential pressure ($P_u - P_d$) at the valve by using two or more valves in series to lower the differential pressure across each valve.
3. Throttle the valve at a different valve opening position by changing the size of the valve. To maintain the same differential and flow rate, a smaller valve may be used at a more open position, or a larger valve may be used at a further closed position. Control valves are often smaller than line size for better control when large energy losses are necessary.
4. Install a smaller bypass line around the main valve to handle low-flow conditions.
5. Install air inlet ports immediately downstream of the valve shaft to admit air and reduce the zone of pressure differential in the pipe. The system must be capable of withstanding air, or provision should be made to remove the air (i.e., incorporating an air release valve or valves). This technique is discussed in other sources (such as Tullis 1989, pages 145–165), and the effects of air induction should be considered in system design or review.
6. Manufacturers may offer cavitation-reducing trim, which can be added to quarter-turn valves.

Table 4-1 Cavitation calculation data for constant-head source 24-in. (600-mm) valve example

Valve Angle θ *(deg.)*	$K_{v\theta}$*	Valve Loss $\Delta H_{v\theta}$ *ft (m)*	ΔH_{sys} Upstream of Valve (75%) *ft (m)* †,‡	$\sigma_{operating}$	PSE	Y	SSE	6-in. Test Incipient Cav. Index σ_i	6-in. Test Constant Cav. Index σ_c	SSE- and PSE-Corrected Incipient Cav. Index σ_i	SSE- and PSE-Corrected Constant Cav. Index σ_c
90	0.30	0.9 (0.29)	74.3 (22.6)	165.79	0.888	0.41	1.754	27.18	14.72	41.77	22.36
80	0.40	1.2 (0.37)	74.1 (22.6)	124.92	0.888	0.38	1.687	19.09	11.32	28.11	16.46
70	1.10	3.4 (1.03)	72.5 (22.1)	46.91	0.891	0.29	1.501	13.21	8.57	17.32	11.12
60	3.10	8.9 (2.72)	68.3 (20.8)	18.15	0.898	0.23	1.368	9.13	6.40	10.98	7.63
50	8.30	20.8 (6.34)	59.4 (18.1)	8.24	0.912	0.18	1.278	6.47	4.75	7.37	5.36
40	24.80	44.0 (13.40)	42.0 (12.8)	4.31	0.938	0.13	1.205	4.83	3.54	5.33	3.87
30	83.30	72.5 (22.10)	20.6 (6.3)	2.92	0.968	0.10	1.148	3.83	2.70	4.14	2.89
20	333.30	91.3 (27.84)	6.5 (2.0)	2.48	0.986	0.07	1.102	3.08	2.18	3.26	2.28
10	3000	99.0 (30.16)	0.8 (0.2)	2.35	0.993	0.04	1.058	2.18	1.90	2.24	1.94

Cav.: cavitation

* 6-in. test data is taken from Tullis 1989, page 140.

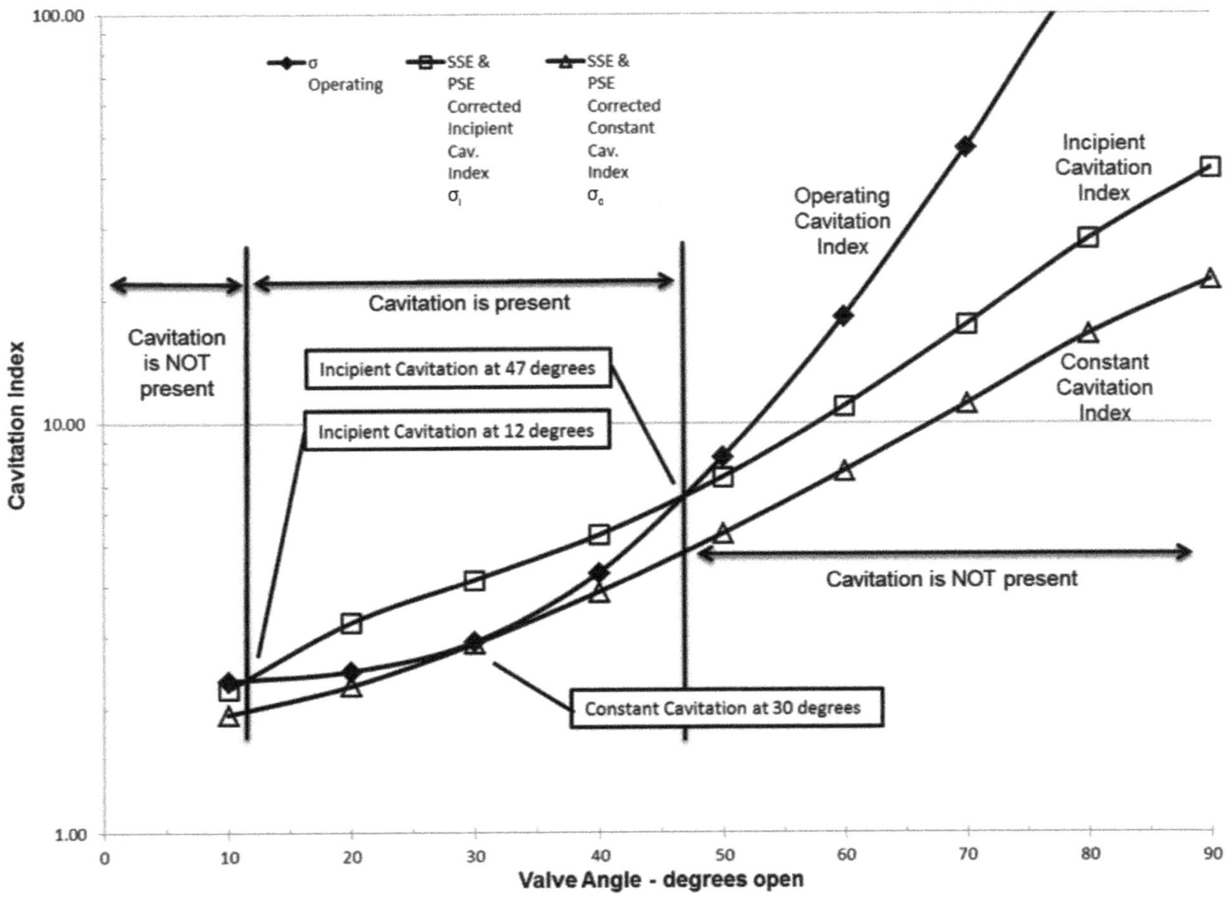

Figure 4-5 Example cavitation analysis summary graph

REFERENCES

ISA (International Society of Automation). 1995. ISA-RP75.23. Considerations for Evaluating Control Valve Cavitation. Research Triangle Park, N.C.: ISA.

Skousen, P.L. 1998. *Valve Handbook*. New York: McGraw-Hill.

Tullis, J.P. 1989. *Hydraulics of Pipelines: Pumps, Valves, Cavitation, Transients*. New York: John Wiley & Sons.

Tullis, J.P. 1993. NUREG/CR-6031. "Cavitation Guide for Control Valves, U.S. Nuclear Regulatory Commission, April 1993." New York: John Wiley & Sons.

This page intentionally blank.

Chapter 5

Valve Testing

The importance of torque, head-loss, and cavitation calculations has been demonstrated. These calculations are only as accurate as the coefficients used in the equations. The purpose of this chapter is to present a practical methodology for testing quarter-turn valves and developing flow, torque, and cavitation coefficients. Testing is recommended but is not a requirement of this manual or any other American National Standards Institute (ANSI)/AWWA standard. Coefficients can also be developed through analytical methods and based on geometric similarities. However, greater accuracy can be expected from testing. Any flow, torque, and cavitation coefficients developed using analytical methods, similarity comparisons, mathematical, or computational fluid dynamic (CFD) models should be verified by testing to determine their accuracy.

UNCERTAINTY

All valves are individuals, just like people. Although we have many common traits that can be anticipated and categorized, we all perform differently in the present and over time. Uncertainty is a complicated subject and not a topic of this manual. Although there is an experimental accuracy associated with any data collection and reduction technique, most laboratories and researchers use measurements of good accuracy and replicated data to minimize the uncertainty in the results. The use of this methodology is intended to provide a best estimate of normal operating head loss, torque requirements, and cavitation potential under the conditions analyzed or specified.

Equipment measurement accuracy is expected to be recorded, and a quality control program is used to assure accurate test results, but specific uncertainty is not designated by this standard practice. The reader is directed to the International Society of Automation (ISA, formerly the Instrument Society of America) and the American Society of Mechanical Engineers (ASME), as both these standards organizations have multiple documents covering this topic. The flow and pressure measurement uncertainty are provided in the following sections of this chapter.

TESTING REQUIREMENTS

The following requirements must be met in designing and conducting flow and torque tests:

1. The test media should be clean water in the range of 35–80°F (10–21°C). Different temperatures normally do not affect head-loss measurements, but variation can affect torque where elastomeric materials are involved. Additional testing is recommended to predict the extent of torque variation with respect to temperature.
2. The upstream and downstream piping should consist of a straight, horizontal run of pipe with the same nominal size as the test valve for a minimum length of 20 times the pipe diameter upstream and 10 times the pipe diameter downstream of the valve. Alternatively, the upstream length of straight piping may be lengthened as required to provide a fully developed uniform, tested, and documented flow profile at a minimum of three times the pipe diameter upstream of the tested valve. Flow conditioners may be used to improve flow conditions in the approach pipe. The piping friction head loss, determined before testing begins, must be subtracted from the measured head loss across the piping run for determining net flow, torque, and cavitation coefficients.
3. Flow, pressure, and torque measurements should be taken at a minimum of 10 positions in the valve's range of travel: typically, 0 (closed), 10, 20, 30, 40, 50, 60, 70, 80, and 90 degrees (open). Many investigators find it best to collect data at 5-degree intervals. Measurements should also include additional data around the peak dynamic torque coefficient position or where a torque direction change (zero transition) occurs. Additional travel position tests are made at the discretion of the researcher, test technician, or manufacturer. Valve positions can be measured with a precision protractor, potentiometer, rotary variable differential transformer (RVDT), or similar device connected directly to the valve shaft. The position reading on the valve actuator should not be used as a reliable indicator of precise position because of hysteresis in the connections and gearing.
4. Model valves may be used for testing, but they must be large enough that their Reynolds numbers exceed 100,000, and they must be geometrically like production valves. Model valves must not be smaller than 6 in. (150 mm) in nominal diameter. The manufacturer must verify the dimensional accuracy of the model to be tested to within 2 percent of actual scaled dimensions.
5. Flow testing determines flow resistance coefficient values, K or K_V; flow coefficient values, C_V; cavitation indexes, σ; and dynamic torque coefficients, C_t. The flow testing must be conducted in accordance with ANSI/ISA S75.02.01, Control Valve Capacity Test Procedure, except that piping manifold losses are subtracted to determine the net K_v for a valve. K values based on the direct measurements, without subtracting the piping manifold losses, will be referred to as gross K_v but will not be used in this methodology.
6. Pressures are measured in the pipe run two times the pipe diameter upstream and six times the pipe diameter downstream of the valve (Figure 5-1). Measurements are taken through static-wall, piezometric pressure taps located on opposite sides of the pipe at each location. The design of the taps must conform to ANSI/ISA S75.02.01, with one exception: For this standard practice, the valve shaft will be mounted vertically so that the effects of closure member weight, buoyancy, and center of gravity do not affect torque measurements and bias the torque data.
7. The volumetric flow is measured with National Institute of Standards and Technology (NIST) traceable weight tanks, volumetric tanks, or other flow sensors (meters or nozzles) within error limits not exceeding 2 percent. The straight length

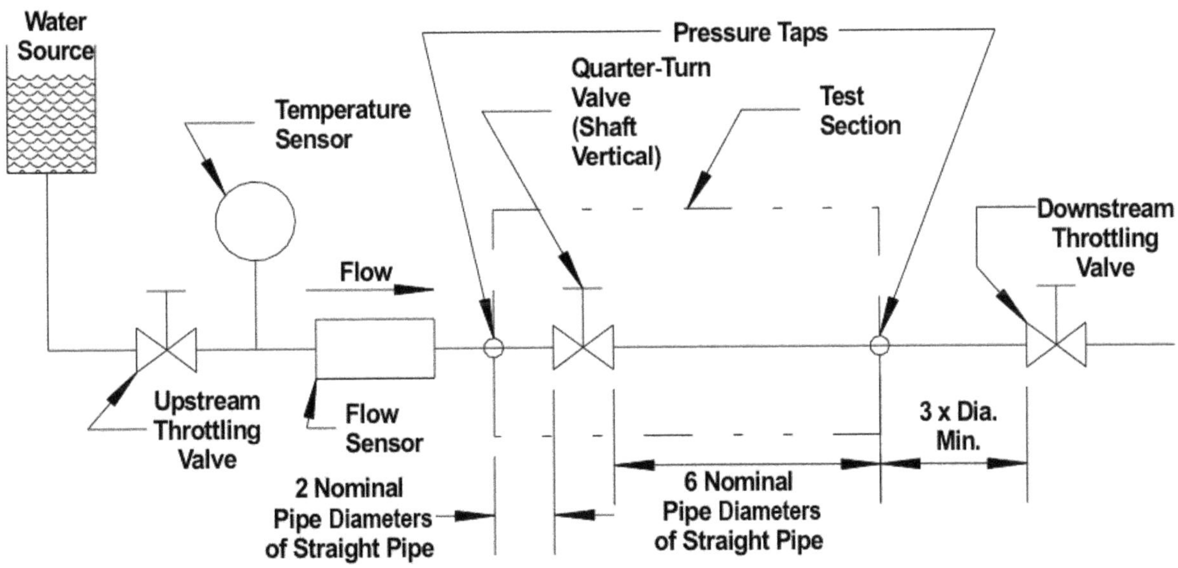

Dia.: diameter

Figure 5-1 Basic flow test system

of pipe upstream and downstream a flow sensor (meter or nozzle) varies by the flow sensor construction but must be sufficient to maintain the 2 percent accuracy. Greater than five (5) pipe diameters is recommended.

8. The accuracy of the pressure measurements must remain within an error range of ±2 percent of the measured pressure differential.

TEST PROCEDURE

The following steps represent a generic procedure for flow testing quarter-turn valves. Because of testing constraints and unusual valve configurations, deviations or anomalies sometimes occur. Such conditions should be explained in the final test report.

Figures 5-1 and 5-2 show the test valve installation. The upstream pressure tap is visible to the right of the valve in Figure 5-2; however, the downstream tap (to the left of the test valve) is not visible. Shaft torque is measured by a strain gauge installed between the actuator and the valve. Flow rate is measured by calibrated flowmeters, weight tanks, or volumetric tanks (not shown).

1. The quarter-turn valve design or scale model should be checked to verify that it has the minimum amount of packing torque on the shaft to provide a reasonable seal. A small amount of packing leakage can be tolerated if it assists in lowering the packing torque, provided it does not affect the flow measurement accuracy or other instrumentation during the test. Rotate the valve in midrange, measure, and record any running, packing, and/or hub seal torque, T_{pt}, with the valve full of water but zero flow.
2. Before installation in the test line, equip the valve with a device to provide a precise indication of valve angle. Mount a strain gauge or torque transducer to the valve shaft to record operating torque. Dynamic torque tending to close the valve should be recorded as positive, resulting in a torque being required by the actuator to open the valve. Dynamic torque tending to open the valve should be recorded as negative, resulting in a torque being required by the actuator to close the valve.

Source: Utah Water Research Laboratory

Figure 5-2 Butterfly valve test installation (NOTE: Flow is from the right to left)

3. The valve should be mounted in the test line with the shaft vertical to avoid the effects of weight and buoyancy torque. This is contrary to the ISA requirement 4.6 regarding pressure tap and shaft location orientations. For nonsymmetrical closure member designs, the flow direction orientation should be recorded. For double- or triple-offset valves, the rotation direction of the lateral offset or eccentricity torque (tending to open or close the valve) should be recorded.
4. The pipe run should be equipped with appropriate flow and pressure measurement devices such as flow tubes, manometers, and pressure transducers or transmitters. The pipe should be pretested to determine the head loss over the eight (two upstream plus six downstream) diameters of the test pipe run.
5. The instrumentation equipment data including any identification numbers, manufacturer, model numbers, scale, accuracy, and calibration last and due dates should be recorded.
6. With the quarter-turn valve at the fully open position, subject the valve to flow in the range of 4–16 ft/s (1.3–5.2 m/s) and record flow and head loss. Repeat the test of fully open flow at a minimum of three different flow rates. Calculate the flow coefficient, K_v, for the valve at each test point. Subtract the pipe head loss from the pressure measurement to obtain ΔH_v. The flow equation is Eq 5-1. Compute the mathematical average of the calculated K_v values and round the result to two decimal places (e.g., 0.32). Repeat the flow coefficient test for lower angles at 10 degree or 10 percent (or smaller) increments.

7. Measure the torques required to rotate the valve shaft in the opening and closing directions at each increment of valve position. Torque readings must be taken with the valve rotating so that bearing torques are measured. A rise in opening torque indicates that dynamic torque is tending to close the valve. Measured torques combine dynamic torque, T_d; bearing torque, T_b; and packing and hub torque, T_p. Calculate the dynamic torque on the basis of Eq 5-2 (NOTE: use torque in in.-lb or N-m only). Repeat the torque coefficient test for the same lower angles at 10 degree or 10 percent (or smaller) increments as performed for the flow coefficient testing. The valve should be operating at or below the constant level of cavitation for the collection of flow coefficient and dynamic torque data. It is preferable that dynamic torque data be collected just below or near the constant level of cavitation.
8. The bearing torque (and the apparent bearing coefficient) at a small closure member angle where the closure member does not interfere with the seat can be determined by calculating the difference of the two torque values (opening and closing) and subtracting the packing and hub torque measured in step 1 (NOTE: use torque in in.-lb or N-m only).
9. The quarter-turn valve should be tested at a maximum of 10-degree intervals to determine incipient, constant, and choking cavitation indexes by adjusting upstream and downstream control valves. The upstream pressure is held constant (typically at 70 psig [481 kPa]), while the flow is increased in small increments until the desired cavitation limits are identified. A graph of the logarithmic accelerometer output versus flow rate or cavitation index is helpful. The slope of this curve normally changes at the points of incipient, constant, and choking cavitation, as shown in Figure 4-2. If holding the upstream pressure constant is not possible, the audible method may be used to identify incipient and constant cavitation, and then the laboratory result can be scaled to a predetermined upstream pressure. Additional information on cavitation testing can be found in ISA-RP75.23, Considerations for Evaluating Control Valve Cavitation.
   - Incipient cavitation is indicated by an intermittent popping noise or increase in vibration above the flow turbulence.
   - Constant cavitation is indicated by a steady noise and vibration increasing at a slower rate. The intensity of the cavitation is the same as incipient cavitation, although the occurrence of the cavitation events is steady.
   - Choking cavitation occurs when the flow rate no longer increases with further opening of the downstream control valve while not increasing the upstream pressure. The choking limit may exceed the flow capability of the test loop. When this occurs, these data may be ignored, or alternate means (such as use of a free discharge test pipe and including the liquid pressure recovery factor) may be used to obtain reasonable results.
10. The flow tests should be repeated with flow oriented in the opposite direction for a valve with an offset or asymmetric closure member.
11. Incipient and constant cavitation coefficients (σ_i and σ_c) should be corrected for size scale effects (SSEs) and pressure scale effects (PSEs) according to Eqs 5-8 and 5-6 and reported in a summary table at the same upstream pressure and size (i.e., 12-in. [300-mm] nominal valve size and 70 psig [481 kPa]) (Tullis 1989). These data should include (1) the SSE-corrected size (12 in. [300 mm]) used, (2) the nominal size of the valve tested, (3) the PSE-corrected pressure (70 psig [482 kPa]), and (4) the vapor pressure of the water during the test. When all data are adjusted to the same size (12 in. [300 mm]) and upstream pressure (70 psig [482 kPa]), the results of different valve styles and designs can be directly compared. If the data are not

SSE or PSE adjusted, the actual size, upstream pressure, and water vapor pressure should be reported.

12. Alternative equations may be used to present and predict cavitation data provided the equations are clearly shown and understood.
13. Summarize the test data for each data point and report at least these results:
    - Valve model and materials
    - Construction drawing and revision level or date
    - Valve angle, in degrees open
    - Flow velocity (based on nominal valve size)
    - Total head loss (measured)
    - Head loss (piping)
    - Net head loss (valve only)
    - Average $K_{v\theta}$, T_p, T_b, $T_{d\theta}$, $C_{t\theta}$, and C_f values
    - Opening and closing torque (consisting of $T_d + T_b + T_p$)
    - $\sigma_{i\theta}$, $\sigma_{c\theta}$, or other cavitation coefficients at the reference pressure (see Figure 4-4)

$$K_v = \frac{2 \times g \times \left(\Delta H_t - \Delta H_{\text{pipe}}\right)}{V^2} \tag{Eq 5-1}$$

Where:

Variable	Definition or Description	Units US Customary (SI—Metric)
g	Gravitational constant Acceleration due to gravity, 32.2 ft/s^2 (9.81 m/s^2)	ft/s^2 (m/s^2)
K_v	Flow resistance coefficient of the valve	dimensionless
V	Velocity of fluid flow	feet per second, fps, or ft/s (meters per second, mps, or m/s)
ΔH_{pipe}	Measured head loss across the pipe during testing without the valve	feet of water (meters of water)
ΔH_t	Measured head loss across the valve and pipe during testing	feet of water (meters of water)

$$T_{dt} = \frac{\text{Opening torque} + \text{Closing torque}}{2} \tag{Eq 5-2}$$

Where:

Variable	Definition or Description	Units US Customary (SI—Metric)
Closing torque	Test-measured torque in the closing direction	in.-lb (N-m)
Opening torque	Test-measured torque in the opening direction	in.-lb (N-m)
T_{dt}	Measured dynamic torque from testing (A positive value indicates a tendency to close the valve.)	in.-lb (N-m)

Calculate the dynamic torque coefficient on the basis of the formula:

$$C_t = \frac{1}{U_{c2}} \times \frac{T_{dt}}{D_d^3 \times \Delta P_t} \qquad \text{(Eq 5-3)}$$

Where:

Variable	Definition or Description	Units US Customary (SI—Metric)
C_t	Coefficient of dynamic torque (positive value tends to close valve), general form	dimensionless
D_d	Disc diameter	in. (mm)
T_{dt}	Measured dynamic torque from testing (A positive value indicates a tendency to close the valve.)	in.-lb (N-m)
U_{c2}	Units' conversion factor: US customary for torque in in.-lb: U_{c2} = 1 in./in. US customary for torque in ft-lb: U_{c2} = 1⁄12 (0.0833) in./ft Metric for torque in N-m: $U_{c2} = 1 \times 10^{-6}$ (0.000001) m/mm	 in./in. (in./ft) (m/mm)
ΔP_t	Measured pressure drop across the disc from testing	psid (kPa)

Calculate the bearing torque on the basis of the formula:

$$T_{bt} = \frac{\text{Opening torque} - \text{Closing torque}}{2} - T_{pt} \qquad \text{(Eq 5-4)}$$

Compute the bearing coefficient of friction:

$$C_f = \frac{1}{U_{c2}} \times \frac{8 \times T_{bt}}{\pi \times D_d^2 \times d_s \times \Delta P_t} \qquad \text{(Eq 5-5)}$$

Where:

Variable	Definition or Description	Units US Customary (SI—Metric)
C_f	Coefficient of friction between the shaft or trunnion and bushing (This value may be obtained from a flow test, engineering handbooks, the bearing manufacturer, or the valve manufacturer.)	dimensionless
Closing torque	Test-measured torque in the closing direction	in.-lb (N-m)
D_d	Disc diameter	in. (mm)
d_s	Shaft diameter	in. (mm)
Opening torque	Test-measured torque in the opening direction	in.-lb (N-m)
T_{bt}	Measured bearing torque from testing	in.-lb (N-m)
T_{pt}	Measured packing and hub torque from testing	in.-lb (N-m)

Variable	Definition or Description	Units US Customary (SI—Metric)
U_{c2}	Units' conversion factor: US customary for torque in in.-lb: U_{c2} = 1 in./in. US customary for torque in ft-lb: U_{c2} = 1⁄12 (0.0833) in./ft Metric for torque in N-m: $U_{c2} = 1 \times 10^{-6}$ (0.000001) m^2/mm^2	in./in. (in./ft) (m^2/mm^2)
ΔP_t	Measured pressure drop across the disc from testing	psid (kPa)

$$\mathrm{PSE} = \left(\frac{P_u - P_v}{P_{ut} - P_{vt}}\right)^{0.28} \tag{Eq 5-6}$$

$$Y = 0.3 \times K_v^{-0.25} \tag{Eq 5-7}$$

$$\mathrm{SSE} = \left(\frac{D}{d_t}\right)^{Y} \tag{Eq 5-8}$$

$$\sigma_i = (\sigma_{it} - 1) \times \mathrm{PSE} \times \mathrm{SSE} + 1 \tag{Eq 5-9}$$

$$\sigma_c = (\sigma_{ct} - 1) \times \mathrm{PSE} \times \mathrm{SSE} + 1 \tag{Eq 5-10}$$

NOTE: See Figure 4-3 for typical graphs of cavitation indexes.

Where:

Variable	Definition or Description	Units US Customary (SI—Metric)
PSE	PSE factor for cavitation analysis	dimensionless
SSE	SSE factor for cavitation analysis	dimensionless
Y	Size scale exponent for cavitation analysis	dimensionless
P_u	Reference upstream pressure for cavitation analysis	psi (kPa)
P_{ut}	Upstream pressure from laboratory test for cavitation analysis	psi (kPa)
P_v	Vapor pressure adjusted for temperature and atmospheric pressure (Example: P_v = –14.4 psig [–99.6 kPa] for water at 60°F [16°C], measured at sea level)	psi (kPa)
P_{vt}	Vapor pressure from laboratory test	psi (kPa)
σ_c	Constant cavitation index at a reference pressure, P_u	dimensionless
σ_{ct}	Constant cavitation index from laboratory testing	dimensionless
σ_i	Incipient cavitation index at a reference pressure, P_u	dimensionless
σ_{it}	Incipient cavitation index from laboratory test	dimensionless

NOTE: Pressures may be gauge or absolute but must be consistent.

SEATING/UNSEATING TEST PROCEDURE

The following steps represent a generic procedure for performing a seating/unseating torque test for a quarter-turn valve. Because of testing constraints and unusual valve configurations, deviations and anomalies sometimes occur. These conditions should be explained in the final test report. Tests should be performed for both flow directions for most valves unless the valve design is completely symmetrical. Valves with offset and asymmetrical closure members should be tested in both directions.

1. The valve model, type, and materials of construction should be recorded. The bearing material and friction coefficient, C_f, or the measured bearing friction torque from step 8 of the preceding flow test will be needed for report calculations. The quarter-turn valve should first undergo a shell and seat leak test to verify proper adjustments of the seat and packing (if applicable).
2. Rotate the valve into a midtravel position and measure any packing and hub seal torque ($T_p + T_h$) with the valve full of water but no flow or pressure. Repeat this measurement three times and average the results.
3. With a blind test head on one flange and water in the valve, slowly close the valve with no differential pressure applied and record the highest total closing (seating) torque ($T_s + T_b + T_p \pm T_{ecc}$). Slowly open the valve and record the highest total opening (unseating) torque ($T_{us} + T_b + T_p \pm T_{ecc}$). Repeat this test three times and average the results for a no-pressure condition. For some valves, this should be done at a low differential pressure to assure the seat tightness is achieved.
4. With a blind test head on one flange, pressurize the valve to its rating. Slowly open the valve and record the highest total opening (unseating) torque ($T_{us} + T_b + T_p + T_h \pm T_{ecc}$). Slowly close the valve with the differential pressure applied and record the highest total closing (seating) torque ($T_s + T_b + T_p + T_h \pm T_{ecc}$). Repeat this test three times and average the results.
5. Repeat step 4 at one or more intermediate pressures to determine the pressure-dependent seating coefficients.
6. Compute the unseating torque (T_{us}) and seating torque (T_s) by subtracting T_b (from calculation or from the flow test) and T_p (including hub seal torque if applicable, measured in step 2) from total measured torque. If the valve design is double or triple offset, the eccentricity torque (T_{ecc}) should be added or subtracted depending on rotation direction.
7. Graphically plot the seating and unseating torques on the ordinate (y-axis) and the differential pressure on the abscissa (x-axis). Perform a linear regression of the data to determine the seating and unseating constant and pressure-dependent coefficients.

Although often based on the higher of the opening or closing torque values, separate seating and unseating coefficients may be determined individually when large differences exist. (Note: Use torque in in.-lb or N-m only.)

Compute the seating coefficient, C_{sc}, and unseating coefficient, C_{usc}, on the basis of the graphical data and linear regression analysis of step 7 and the following equations. If the seating and/or unseating torque is reasonably constant at all pressure, then only determining the pressure-independent coefficients is necessary. If these data show a definite positive or negative slope relative to the operating pressure, then the pressure-dependent coefficients should be developed.

The following example and equations are based on the ball valve (BV) and butterfly valve (BFV) seating and unseating methods (Eq 3-5) shown in Chapter 3. For double- and triple-offset BFVs, the alternate coefficients may be developed consistent with the nomenclature provided in Eq 3-6. For plug valves (PVs), the coefficients need to be developed consistent with the nomenclature provided in Eq 3-7.

$$C_{sc} = \frac{1}{U_{c2}} \frac{T_{st}}{D_d^2} \quad \text{(Eq 5-11)}$$

$$C_{usc} = \frac{1}{U_{c2}} \frac{T_{ust}}{D_d^2} \quad \text{(Eq 5-12)}$$

Where:

Variable	Definition or Description	Units US Customary (SI—Metric)
C_{sc}	Constant or pressure-independent coefficient of seating torque	lb/in. (N/m)
C_{usc}	Constant or pressure-independent coefficient of unseating torque	lb/in. (N/m)
D_d	Disc diameter	in. (m)
T_{st}	Measured seating torque (always positive)	in.-lb (N-m)
T_{ust}	Measured unseating torque (always positive)	in.-lb (N-m)
U_{c2}	Units' conversion factor: US customary for torque in in.-lb: U_{c2} = 1 in./in. US customary for torque in ft-lb: U_{c2} = 1⁄12 (0.0833) in./ft Metric for torque in N-m: $U_{c2} = 1 \times 10^{-6}$ (0.000001) m/mm	in./in. (in./ft) (m/mm)

$$T_s = \left(C_{sc} + C_{sp} \times \Delta P_v\right) \times D_d^2 \quad \text{(Eq 5-13)}$$

and:

$$T_{us} = \left(C_{usc} + C_{usp} \times \Delta P_v\right) \times D_d^2 \quad \text{(Eq 5-14)}$$

Where:

Variable	Definition or Description	Units US Customary (SI—Metric)
C_{sc}	Constant or pressure-independent coefficient of seating torque	lb/in. (N/m)
C_{sp}	Pressure-dependent coefficient of seating torque	lb/in./psi (N/m/kPa)
C_{usc}	Constant or pressure-independent coefficient of unseating torque	lb/in. (N/m)
C_{usp}	Pressure-dependent coefficient of unseating torque	lb/in./psi (N/m/kPa)
D_d	Disc diameter	in. (m)
T_s	Seating torque (always positive)	in.-lb (or ft-lb) (N-m)
T_{us}	Unseating torque (always positive)	in.-lb (or ft-lb) (N-m)
ΔP_v	Pressure drop (or loss) across the valve, general form	psid (kPa)

CHOKED FLOW TESTING

The testing for liquid pressure recovery factor of a valve without attached fittings ($F_{l\theta}$, $F_{l\theta}^2$) typically involves preforming multiple flow tests at increasing flow rates at the same inlet pressure to the valve to observe the onset of flow rate choking. This testing is performed less often for valves in this scope, as these are not typically used in critical flow-control service. For this testing, refer to ANSI/ISA S75.02.01 and ISA-RP75.23, which address the testing and use of these flow factors. Eq 2-21 is used as the basis of this evaluation.

REFERENCES

ANSI (American National Standards Institute)/AWWA. 2015. C504. Rubber-Seated Butterfly Valves. Denver: AWWA.

ANSI/ISA (International Society of Automation). 2008. S75.02.01-2008. Control Valve Capacity Test Procedure. Research Triangle Park, N.C.: ISA.

ISA. 1995. ISA-RP75.23-1995. Considerations for Evaluating Control Valve Cavitation. Research Triangle Park, N.C.: ISA.

Tullis, J.P. 1989. *Hydraulics of Pipelines: Pumps, Valves, Cavitation, Transients*. New York: John Wiley & Sons.

This page intentionally blank.

Chapter 6

Valve Applications

This chapter provides recommendations for actuator sizing and valve installations. Some piping configurations encountered in water systems can dramatically affect head loss through a valve and its operation. These effects should be understood by the system designer. The chapter also includes cautions that should be observed when quarter-turn valves are used for throttling service, when they are subject to unusual upstream flow conditions, and when actuators are removed.

ACTUATOR SIZING

The formulas for determining total break, seating, unseating, opening, and closing torques are presented in Chapter 3. Actuator sizing should be based on a comparison of the highest torque values for the valve with the torque rating and output capability of the actuator. The Chapter 3 Torque Components section total operating torque is the "minimum required shaft torque" (MRST) at each travel position. This torque is the most probable torque requirement for the valve when operating under the system conditions analyzed.

The torque formulas do not include application factors (AFs) that should be considered when sizing powered actuators. Refer to the applicable AWWA valve standard for AFs and other considerations necessary to properly size manual, cylinder, and electric actuators. The MRST values are multiplied by the appropriate AFs given in these standards to determine the "actuator sizing torque" (AST = MRST × AF) for sizing powered actuator driver margins. In the AWWA standards, the authors have used the term "AFs" because there are different values associated with the actuator type and application function. Other industry standards use the term *safety factor* or *safety margin* in a similar manner.

Manual Actuator Sizing

Manual actuators are sized on the basis of two criteria. First, the actuator rating must exceed the MRST at all valve positions. Second, the actuator must be sized to allow operation of

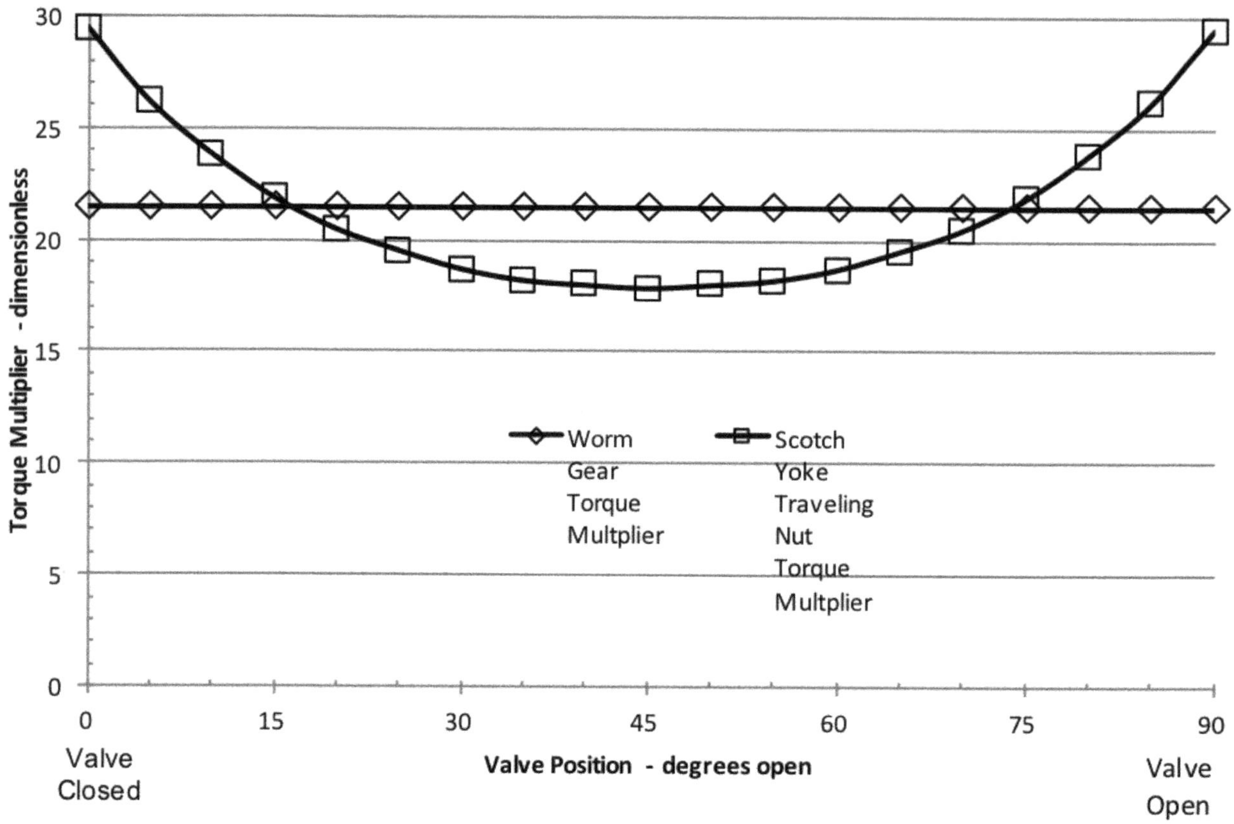

Figure 6-1 Typical actuator torque characteristics

the valve without exceeding certain limits on handwheel or chainwheel pull force or input torque (such as 80-lb [356-N] handwheel rim pull force or 150 ft-lb [219 N-m] input torque of the current American National Standards Institute [ANSI]/AWWA C504, Rubber-Seated Butterfly Valves). The calculation of rim pull force or input torque requires a review of the characteristic torque curve for the actuator (see Figure 6-1).

The typical AWWA maximum handwheel or chainwheel pull requirement of 80 lb (36.3 kg) has been found by some operating personnel to be a high exertion of effort, and lesser pulls of 40–60 lb (18.1–24.2 kg) have sometimes been found to be beneficial. This may require more input turns, a larger handwheel or chainwheel, or perhaps larger or more efficient actuators.

Typical torque multiplier curves are presented for both a worm-gear actuator and a traveling-nut–type actuator in Figure 6-1. The torque multiplier is the expected ratio of output torque to input torque, considering the efficiency of the mechanism and/or gearing mechanical advantage. For example, if an actuator has a torque multiplier of 20 and a butterfly valve (BFV) requires an operating torque of 8,000 in.-lb (666 ft-lb, 973 N-m), then the actuator will require an input torque of 8,000 in.-lb/20 (666 ft-lb/20, 973 N-m/20) or 400 in.-lb (33.3 ft-lb, 49 N-m).

The worm-gear actuator has a constant torque multiplier throughout the full travel, and the "curve" in Figure 6-1 is a horizontal straight line at all valve opening positions. Conversely, the traveling-nut–type actuator, which includes scotch yoke or link and lever designs, has variable torque multipliers depending on valve opening position. When calculating the input torque requirements of an actuator with a variable torque

Figure 6-2 Worm-gear actuator cutaway

characteristic curve, use the operating torque at each valve opening position and the corresponding actuator torque multiplier.

The input torque is calculated by dividing the required valve torque by the actuator's torque multiplier. Handwheel rim pull force is calculated by further dividing the input torque by the radius of the handwheel.

Example cutaway images of a worm-gear actuator (Figure 6-2), a scotch yoke traveling nut actuator (Figure 6-3), and a link and lever traveling nut actuator (Figure 6-4) are provided, which display the internal mechanisms. The scotch yoke and the link and lever traveling nut actuators are often used on quarter-turn valves, as the torque multiplier or mechanical advantage is greatest at the closed position. In smaller valves, this is advantageous because the seating and unseating torque is often the greatest torque requirement. In larger valves, where the greatest operating torque may be at midstroke positions, this is still an advantageous characteristic because this slows the rotation speed as the valve approaches the closed position and assists in reducing the risk of a pressure transient caused by too fast a valve closure.

Cylinder Actuator Sizing

Five basic types of pneumatic or hydraulic actuators are used on AWWA quarter-turn valves. These include rack and pinion, scotch yoke, link and lever, pivoting cylinder, and vane. The first four types require a separate mechanism (gear or lever) to convert the

Figure 6-3 Scotch yoke traveling nut actuator cutaway

Figure 6-4 Link and lever traveling nut actuator cutaway

cylinder's linear thrust output into torque for operating the quarter-turn valve. The vane actuator is the only one that converts fluid pressure directly into torque.

The mechanical gear or lever portion of a cylinder actuator should be designed to handle the total torque (MRST) calculated for manual actuators. Selection of the cylinder bore size (or area) should be based on the minimum supply pressure to the cylinder and the AST. Additionally, ANSI/AWWA C504; ANSI/AWWA C507, Ball Valves, 6 in. Through 60 in. (150 mm through 1,500 mm); ANSI/AWWA C516, Large Diameter Rubber-Seated

Butterfly Valves, 78 in. (2,000 mm) and Larger; and future ANSI/AWWA C517, Resilient-Seated Cast-Iron Eccentric Plug Valves, recommend AFs based on the type of cylinder controls when sizing the actuator. (NOTE: The current edition of ANSI/AWWA C517 does not use the term AF. The newer editions are anticipated to conform to this nomenclature.)

The AF is needed to provide sufficient input power and allow for pressure drop in cylinder control valves and speed control devices. Higher AFs are used for cylinders powered by air and for throttling flow because the cylinder is moved by a floating differential pressure across the piston, which is created by a small orifice in the positioner. Because of the compressibility of the air, cylinder throttling without a positioner should be avoided.

Some cylinder actuators are equipped with springs for fail-safe closure or fail-safe opening. A spring-return cylinder requires more analysis of sizing based on calculated valve required operating torques, cylinder AFs, and the variable torque generated by the spring. In all cases, the actuator manufacturer's data and sizing information should be consulted for proper actuator sizing.

Many cylinder actuators have a nonlinear variable output characteristic relative to valve position. When this occurs, the actuators' AFs (or margins) should be evaluated at each calculated valve position as illustrated in Figure 6-5.

Refer to standard ANSI/AWWA C541, Hydraulic and Pneumatic Cylinder and Vane-Type Actuators for Valves and Slide Gates, for the design and construction of these actuators. Appendix A of ANSI/AWWA C541 has a useful actuator data form for selecting and specifying actuator requirements.

Electric Actuator Sizing and Switch Settings

The gear portion of the electric actuator is sized on the basis of the total break and running torques (MRST) provided by this manual of practice. Additionally, the motor should be sized on the basis of the minimum expected motor voltage plus the AFs recommended

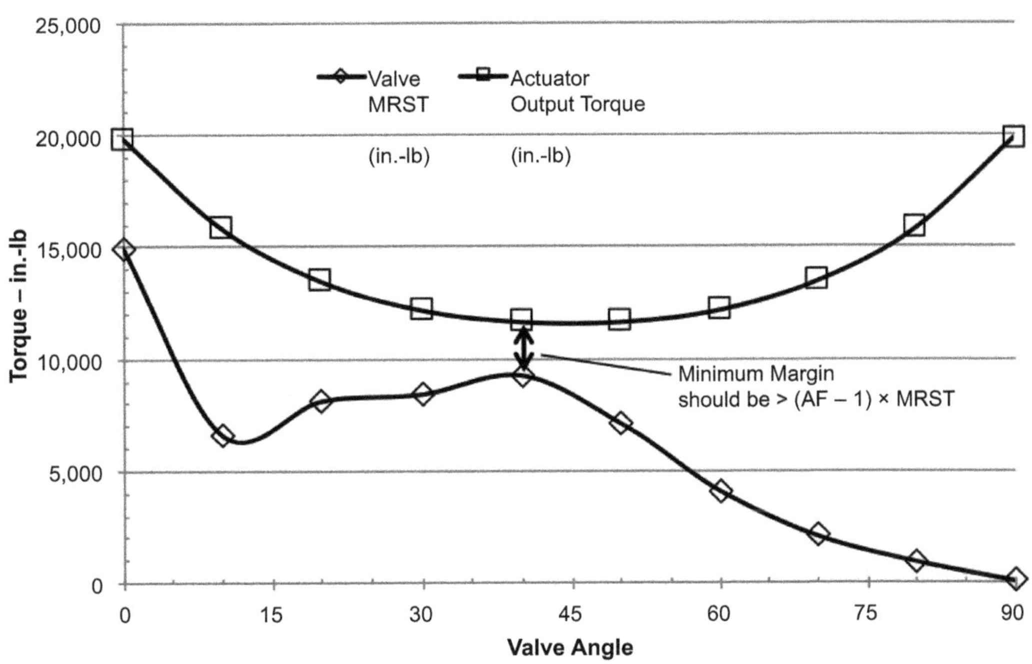

Figure 6-5 Actuator sizing characteristics graph

by ANSI/AWWA C504; ANSI/AWWA C507; ANSI/AWWA C516; ANSI/AWWA C517; or ANSI/AWWA C519, High-Performance Waterworks Butterfly Valves—3 In. (75 mm) Through 60 In. (1,500 mm).

An electric actuator for a quarter-turn valve may be wired for limit switch position seating or torque seating as indicated by the manufacturer. Symmetric and single-offset rubber-seated valves are usually position seated, whereas double- and triple-offset valves may be torque seated. For position-seated valves, the torque switch is used for over-torque protection. It can then be set above the AST and below the actuators' rating or the torque, which stalls the motor. This protects the unit from burnout should an unusually high torque be encountered (such as an obstruction in the pipeline).

Refer to standard ANSI/AWWA C542, Electric Motor Actuators for Valves and Slide Gates, for further design and construction requirements of these actuators.

EXTENDED BONNET INSTALLATION

When the quarter-turn valve must be operated from a significant distance above because of a structural, buried, or submerged condition, the valve may be equipped with an extended bonnet. An extended bonnet consists of an outer pipe that is rigidly attached to the valve body and extends up and rigidly supports the actuator. The valve shaft extends up through the extended bonnet with a shaft or inner pipe and connects to the actuator mechanism. The inner pipe then rotates 90 degrees to operate the quarter-turn valve. The typical construction is illustrated in Figure 6-6.

To assure tight shutoff of the seated quarter-turn valve, the actuator and extended bonnet must position the valve closure member to within a limited angle depending on the valve design. For example, ball valves (BVs) and BFVs typically need to position the closure member to within about ±1 degree of the true closed position. Since plug valves (PVs) have eccentric action and may be torque seated, the angle may be higher, typically ±2–3 degrees. Because both the outer and inner pipes see the full operating torque of the valve, they are subject to torsional deflection. The total torsional deflection is the sum of the deflections of the outer and inner pipes. As the length of the bonnet increases, the deflection increases proportionally. Hence, it is often more important to size the extended bonnet pipes to limit the torsional deflection rather than strength and allowable stress. During the midstroke positions, the torsional deflection is not as important, and stress becomes more critical. However, any torsional deflection at midstroke positions should be limited to about 3–5 degrees.

With longer extensions (greater than approximately 6 ft [1.8 m]), to limit the torsional deflection to low values may become impractical because the required bonnet pipe sizes become exceptionally large. As an alternative, some extended bonnet designs may have a closed stop limiting bolt near the valve. When a closed position stop is used, the extended bonnet is allowed to deflect more, and the closed stop precisely positions the valve disc in the closed position. In this case, the extended bonnet can be allowed to deflect up to 3 or 5 degrees, depending on the allowable over-travel in the actuator. Whether the valve is position seated or torque seated, it is good practice to base the design of extended bonnets on deflection requirements as well as material strength requirements.

EFFECTS OF PIPE INSTALLATIONS

Proper installation can prevent serious problems with valve performance and life expectancy. Many operating conditions should be reviewed with the manufacturer, such as flow rate, differential pressure, and temperature. A comprehensive list of information that should be included when placing orders is given in the forewords of ANSI/AWWA

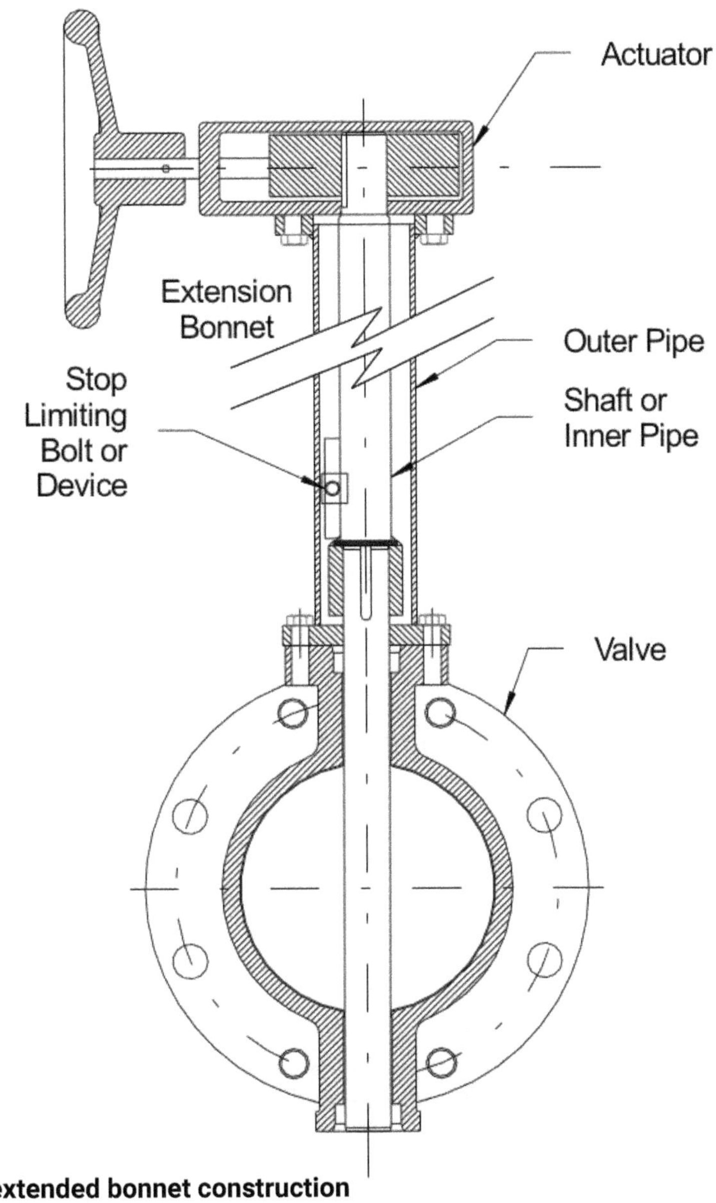

Figure 6-6 Typical extended bonnet construction

C504, ANSI/AWWA C507, ANSI/AWWA C516, ANSI/AWWA C517, ANSI/AWWA C519, and ANSI/AWWA C522. In addition, the following recommendations should be followed in the placement and installation of a quarter-turn valve in a piping system. Appendix A is a suggested specification and data sheet that can be used to communicate quarter-turn valve design requirements in accordance with the AWWA quarter-turn valve standards.

The design of quarter-turn valves is based on straight pipe testing in laboratory conditions. Any upstream piping components that cause asymmetric flow conditions or increased turbulence at the valve will alter the valve's performance as well as the system's performance. In many cases, these changes are benign, but several installation conditions can severely alter quarter-turn valve performance. The following examples provide some basic guidelines for the use and installation of quarter-turn valves. Although many of these figures depict a BFV, the same effects apply to all quarter-turn valves.

Flow and Pressure Direction

BFVs with symmetrical discs can typically be installed with flow and pressure in either direction. BFVs and PVs with asymmetric closure members have different flow as well as seating, unseating, and dynamic torque properties depending on whether the shaft is upstream or downstream of the seat. Also, many asymmetric valves tend to seal better with the shaft on the upstream side of the valve seat (review Figures 3-1 and 3-3). A valve may have a preferred orientation based on the pressure assisting the seal performance, but the seating, unseating, and dynamic torque requirements may require the opposite orientation. Check with the manufacturer before installing a valve for the preferred direction of flow or pressure if not indicated on the valve body or in their instruction manuals or drawings.

If system configuration or operating modes require a valve to be installed opposite to the valve's preferred installation orientation, consult the manufacturer for recommendations or modifications.

The valve installation can also be affected by the actuator configuration. Conditions may favor orienting the handwheel or operating nut in a specific direction. If the desired actuator orientation does not match the required flow direction and system orientation, then the valve manufacturer should be consulted about reorienting an actuator.

General Upstream Flow Disturbance

The first consideration is the flow rates. If the flow rate and corresponding fluid velocity are low, then the effect of an upstream flow disturbance is likely inconsequential to the valve's dynamic performance (head loss and torque). As the local velocities that approach the valve become greater, their effect on the dynamic torque and head loss increases, and atypical loads and stresses are developed in the valve. The severity of the discontinuity in the local velocity patterns at the valve increases the uncertainty in the valve's performance. Some research has been done on the effects of elbows and pumps upstream of quarter-turn valves, but the data are limited because there are many variables to consider. These variables include valve type, elbow or pump orientation relative to the valve shaft axis, distance between the disturbance and the valve, closure member rotation direction, and fluid velocity (or flow rate).

These conditions affect the fluid dynamic responses of the valve (head loss and dynamic torque). For valves that have low friction–related torque components, these dynamic effects can be of greater consequence than for valves that have high friction–based torque requirements. As an example, a valve using PTFE packing, a clearance seat, and roller bearings will have little operating friction torque requirements, and any change in the dynamic torque will be of major significance to the total operating torque. Conversely, a valve constructed of asbestos-free braided pull-down packing, low-leakage metal seats, and stainless-steel bearings and shafts will have significant operating friction torque, and the changes in dynamic torque will not greatly affect the total operating torque requirement.

For most water works applications, some general rules of thumb can be used provided that the maximum velocities are less than the AWWA Class "B" velocity designation of about 16 fps. For best dynamic performance (head loss and torque), a length of straight pipe between the disturbance and the valve greater than eight times the pipe diameter is sufficient to provide normal flow through the valve. For good performance, there should be two or three pipe diameters of straight pipe upstream of the valve. Less than two pipe diameters of upstream straight pipe may impair valve performance at higher fluid velocities. Dynamic torque can be doubled or tripled by an improperly oriented valve installation. The valve shaft should be oriented such that the fluid flow on either side of the shaft centerline is essentially equal.

When operating conditions are extreme or operation and function are critical, reference a similar operating experience or perform model testing of the installation.

Upstream Elbow or Branch Tee

Elbows and branch tees cause asymmetrical fluid velocity in the pipe (Figure 6-7), which affects quarter-turn valve operation. This figure and others in this section depict BFVs, as these are generally more vivid visual representations. However, these principles apply

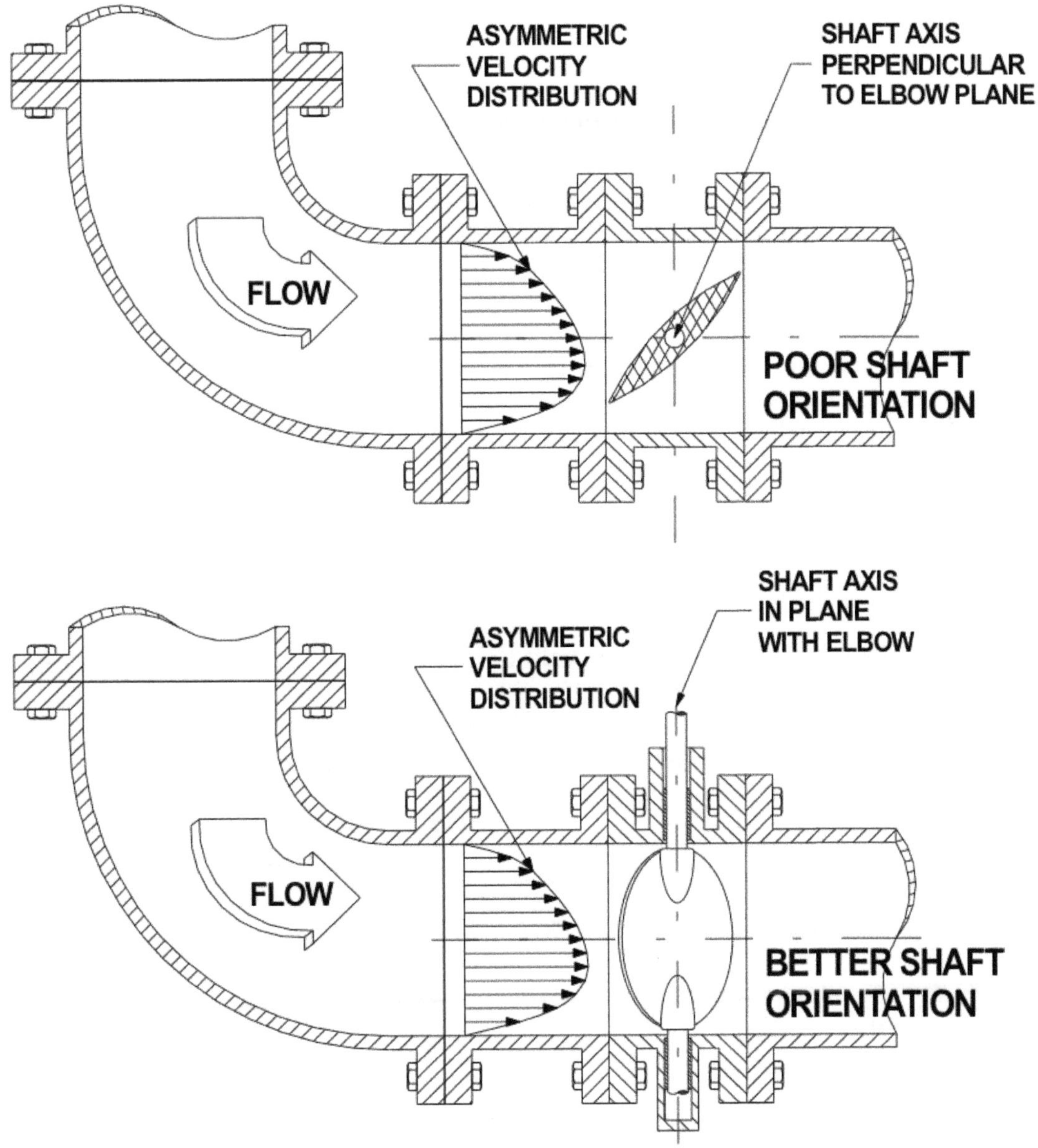

Figure 6-7 Vertical elbow upstream of a BFV, or other quarter-turn valve, preferred installation orientation

to all quarter-turn valve installations. Dynamic torque can increase significantly by an improperly oriented valve and an upstream elbow. The valve shaft must be positioned vertically when installed downstream of a vertical elbow or tee. For a horizontal elbow or tee, the valve shaft should be positioned horizontally.

Upstream Expansion

Expansion sections cause major fluid pattern changes and turbulence as the flow approaches a downstream quarter-turn valve. When short expander laying lengths are used and close-coupled, the disc of BFVs may interfere with the expander inside diameter (ID) when operated. Expanders upstream of quarter-turn valves should be concentric rather than eccentric whenever possible. Similar to the upstream elbows, the distance upstream should be two or more diameters for good performance. If an eccentric arrangement is necessary, the shaft axis should be oriented in the direction of the offset as shown in the lowest orientations of Figure 6-8. Figure 6-8 illustrates preferences and compatibility for several orientation arrangements. Although depicted with BFVs, these scenarios apply to all quarter-turn valve installations.

Upstream Orifice and Venturi Flowmeters

Upstream orifices also cause major fluid pattern changes and turbulence as the flow approaches a downstream quarter-turn valve. Orifices upstream of quarter-turn valves should be concentric rather than eccentric whenever possible. Similar to the upstream elbows and expansions, the distance upstream should be two or more diameters for good performance of both the quarter-turn valve and the orifice. In this case, the orifice performance will be altered as well as the quarter-turn valve performance if there is insufficient spacing. This is especially true when the orifice is used as a meter. A close-coupled control valve can alter the pressure tap reading and affect the accuracy of the meter. Venturi meters, which use upstream and throat pressure taps, are unlikely to be affected by

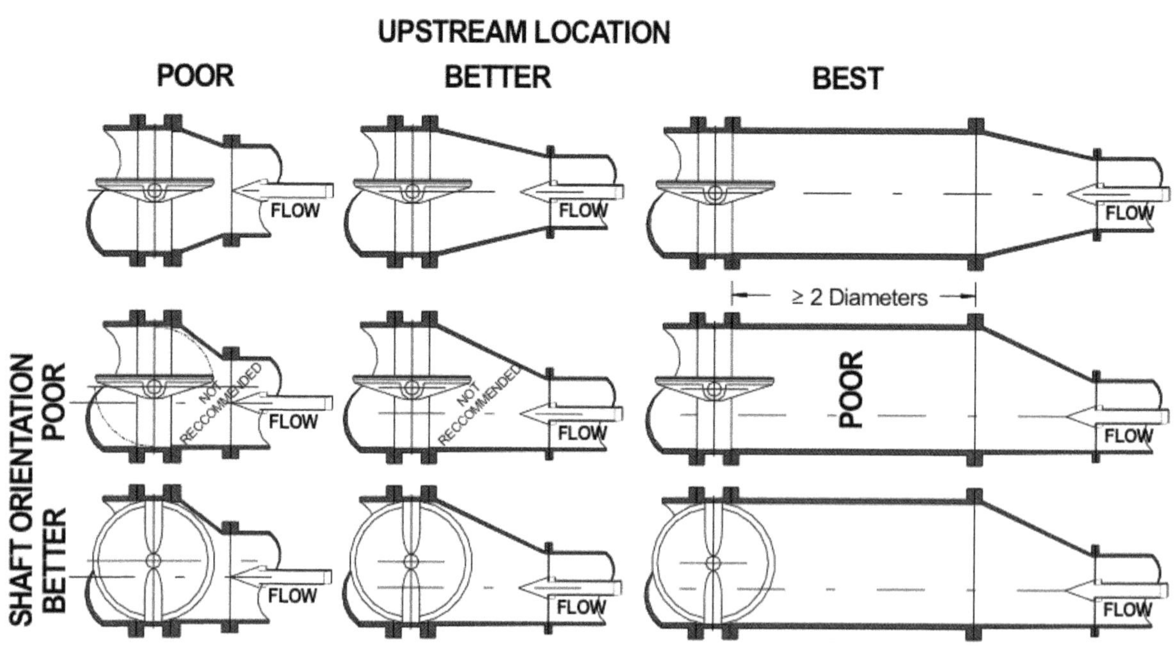

Figure 6-8 Upstream expansion orientation preferences

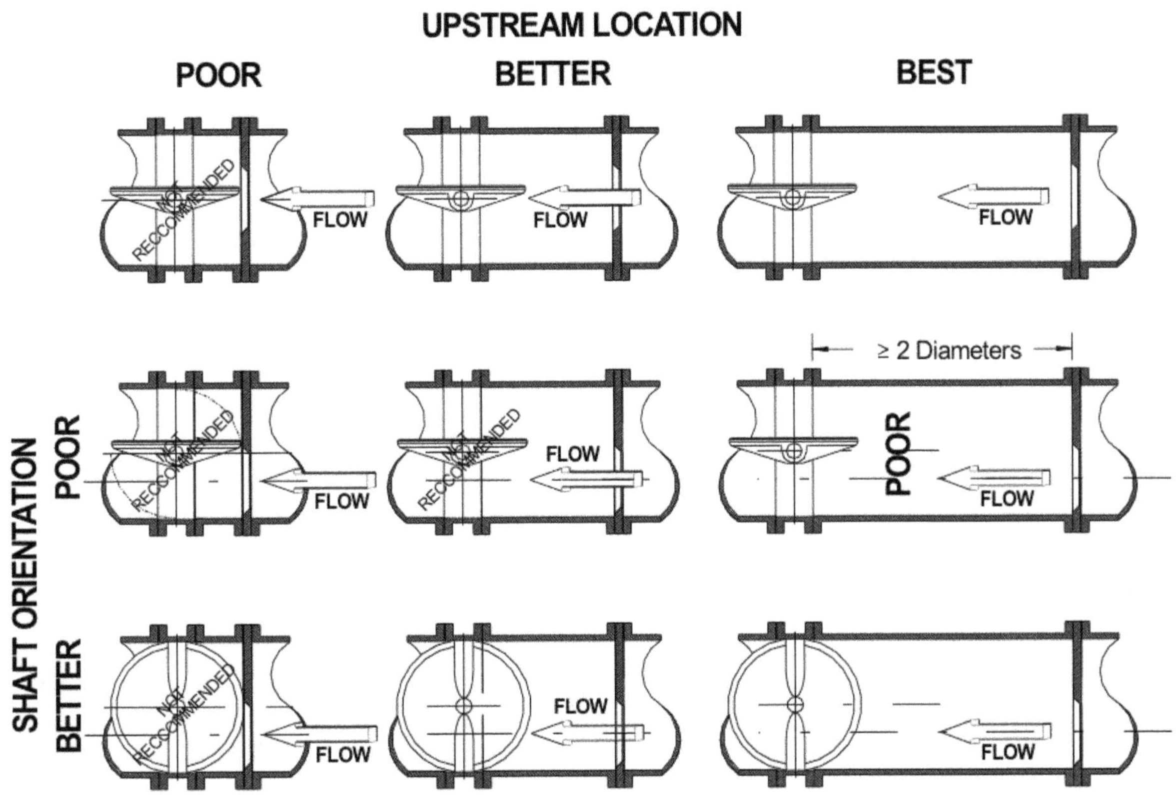

Figure 6-9 Upstream orifice orientation preferences

a close-coupled downstream control valve. However, venturi meters, whether full-length flanged or insert, often have reduced exit diameters and can interfere with valve closure member operation if not sufficiently separated.

If an eccentric orifice arrangement is necessary, the shaft axis should be oriented in the direction of the offset as indicated in the lowest line of Figure 6-9. See Figure 6-9 for additional preferences and compatibility. Again, these depictions with BFVs apply to all quarter-turn valve installations.

Upstream Reducers

Upstream reducers can also cause poor valve behavior if not oriented properly. Concentric reducers are best, and again, it is better to maintain a greater distance upstream for good performance. As with the other orientations, the approach flow profile should be essentially equal on either side of the shaft axis. Again, these depictions with BFVs apply to all quarter-turn valve installations.

If an eccentric upstream reducer arrangement is necessary, the shaft axis should be oriented in the direction of the offset as indicated in the lowest line of Figure 6-10. See Figure 6-10 for additional preferences and compatibility.

Upstream Valves

A common type of valve positioned upstream of a quarter-turn valve is a check valve (Figure 6-11). Less than one diameter of straight pipe between the check valve and the quarter-turn valve will cause these valves to interact. When the check valve has a horizontal

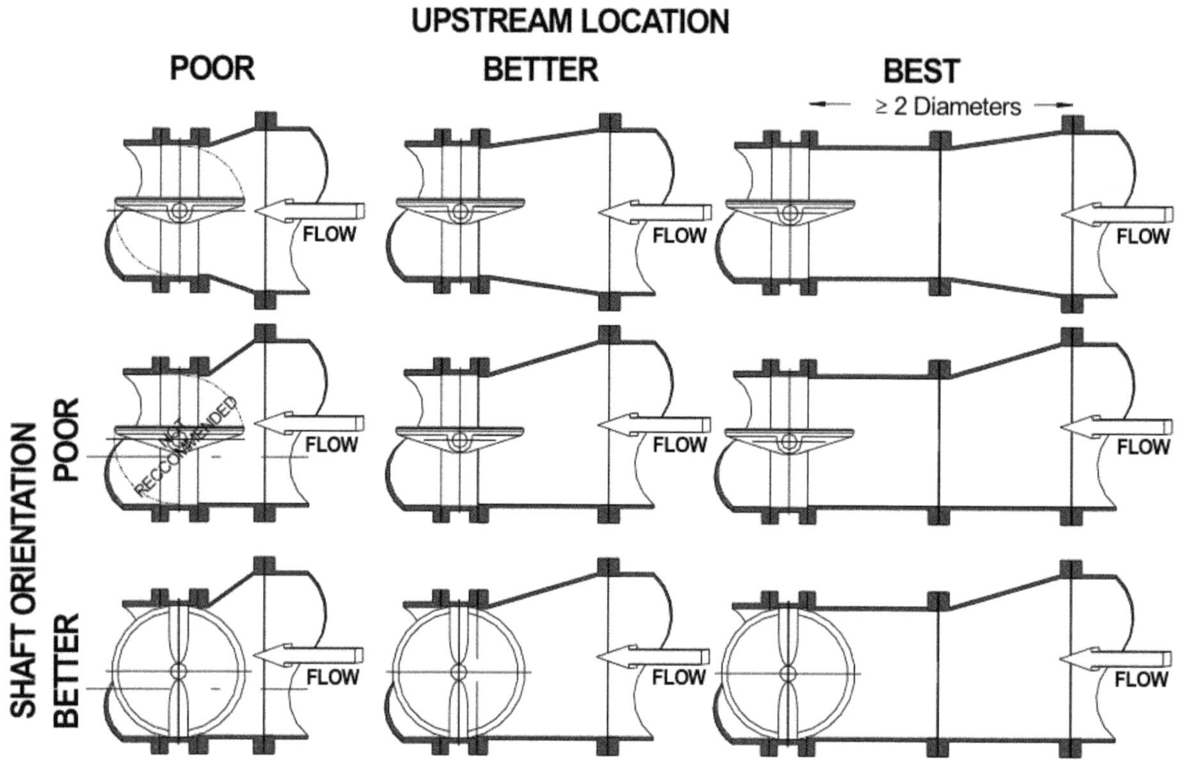

Figure 6-10 Upstream reducer orientation preferences

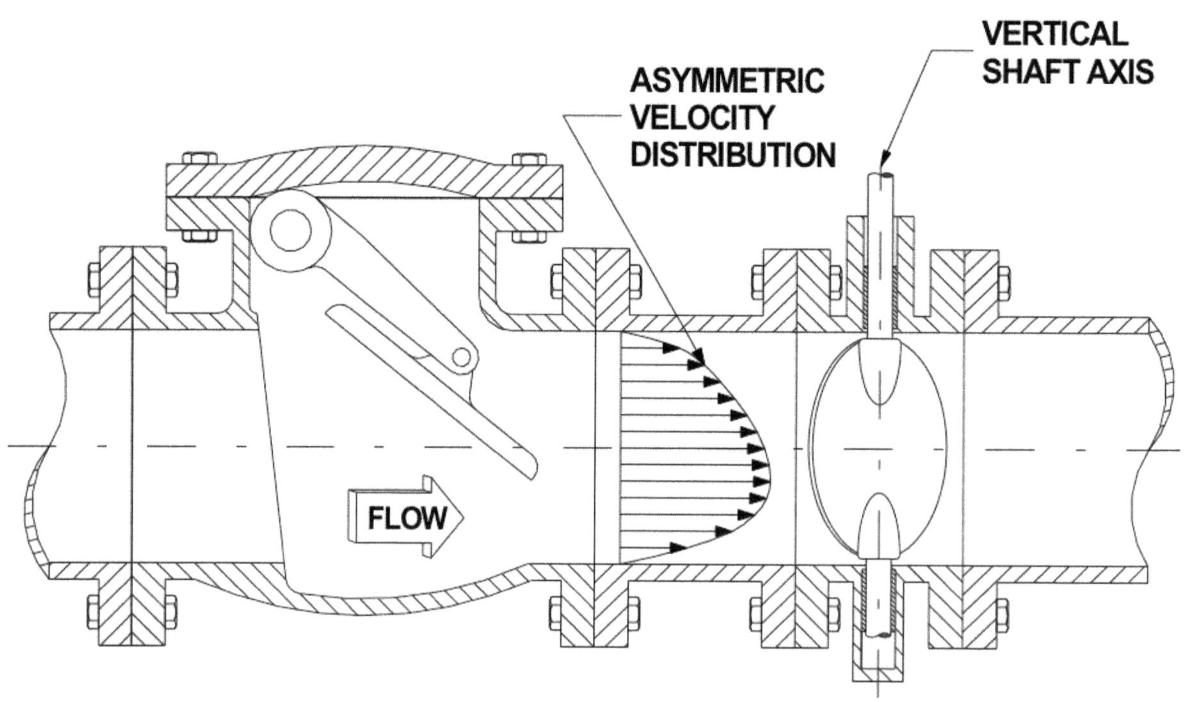

Figure 6-11 Upstream check valve preferred installation orientation

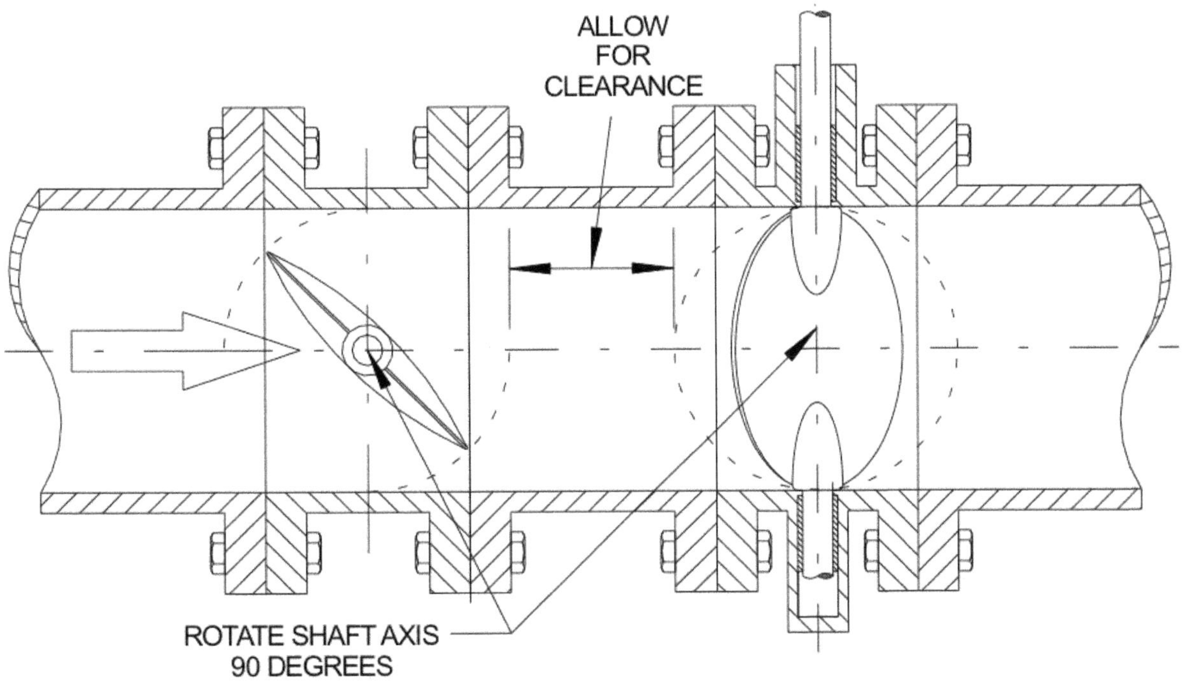

Figure 6-12 Series mounted valve preferred installation orientation

pivot shaft (or hinge pin), the quarter-turn valve should be installed with a vertical shaft. This positioning allows the high localized fluid velocities to be divided evenly across the quarter-turn valve's shaft. When two quarter-turn valves are placed in close proximity (less than three diameters), their shafts should be perpendicular to each other so that the upstream valve does not cause excessive torques in the downstream valve. This orientation is demonstrated in Figure 6-12 and is relevant to all quarter-turn valves.

Many valves and other piping components have internal parts or closure members that may extend beyond the end connection during operation. Series mounted valves and other components with internal parts or closure members should be placed far enough apart to avoid any structural interference. For instance, the centerlines of two series piped BFVs should be at least one diameter apart so that their discs do not touch during operation. These valves would function better with two or three diameters between centers to reduce the possibility of fluid dynamic interaction.

Furthermore, never combine two or more poor orientations such as the following: (1) short expansion, (2) improperly oriented shafts, and (3) highly throttled upstream valve illustrated in Figure 6-13.

Free Inlet or Discharge

Quarter-turn valves are often mounted on the walls or partitions of basins and tanks, as in Figure 1-7. The valve manufacturer should be aware of such an installation. Special torque and flow coefficients may be used to evaluate these conditions. Also, the direction of flow may be more critical for an asymmetric valve. The valve shaft is usually mounted in the vertical orientation so that partial flows are equally divided across both sides of the shaft axis.

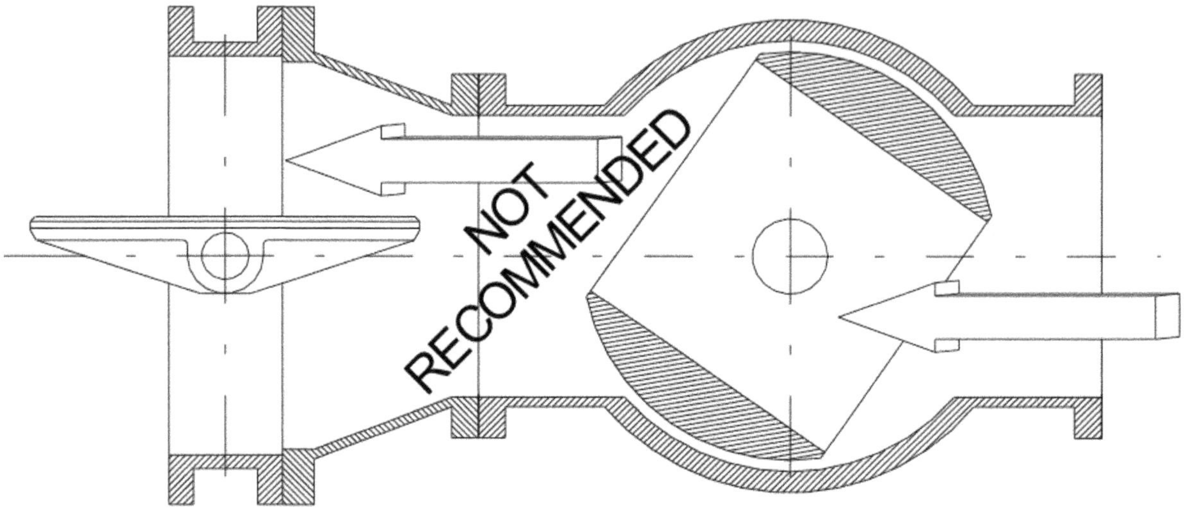

Figure 6-13 Combined improper orientations

External and Pipe Loads

Quarter-turn valves are self-contained devices that may not function properly or remain tight if subjected to external forces. Quarter-turn valves are not designed or intended to support adjacent piping and should not be used to transmit piping loads across the body of the valve. Quarter-turn valves are designed to withstand pressure vessel and normal operating loads and to be stronger than the adjacent piping. Additional external and piping applied loads may affect valve performance and operation. Quarter-turn valves rely on a precise fit between the body and closure member seats to form a low-leakage or leak-tight seal. Although valve bodies are designed to be stronger than the mating pipe under internally applied loads and stresses, they are still susceptible to deflection or distortion from externally imposed and applied piping loads. Any valve deflection or distortion caused by these loads impairs the valve's ability to properly seat and create a positive seal. Even if deflection or distortion does not cause the leakage rate to be observably increased, the valve's service life and operating ability could be affected.

If a valve is rigidly installed in a pipeline using flanged joints, the whole assembly of pipe and valve can be stressed by temperature changes, settlement, pipe support, and (if buried) exceptional surface loads. To avoid a valve being overly externally stressed, strained, and/or distorted, there should be at least one flexible joint installed next to or near the valve. Never connect a flanged valve to an out-of-tolerance, out-of-round pipe or flange. Refer to the flange standards for the bolting dimensional tolerances. The connection and support structure next to the valve shall prevent or at least minimize the piping loads and bending moments from being transmitted through the valve body to stop the potential for overstressing the valve flanges and deflecting the body. Having two flexible couplings near and one on either side of a valve is best if the section between the couplings can be restrained and simply supported reducing or relieving the transmitted pipe loads to the valve.

For buried applications of large-diameter flanged valves, the designer should consider all pipe, valve, soil, and surface live loads in designing the supports for the adjacent piping. The designer may consider using a common foundation slab and concrete or steel pipe cradles to support the adjacent piping, which could include the use of flexible pipe joints on one or both sides of the valve to ensure that any piping or external loads do not deflect or distort the valve's body. It is the responsibility of the system or piping designer to determine the loading conditions and the required valve and pipe supports for each specific installation. While these and many other construction methods are suitable, the system and piping designer must understand that pipe loads transmitted through the valve causing even a small distortion of the valve seat may affect the valve's ability to properly seal, change the operating torque requirements, alter system fluid flow characteristics, and reduce the valve's service life.

Buried valves normally should use mechanical joint-type restraint fittings that allow for some flexibility and relief of the piping loads. These joints reduce the pipe loads transferred to the valve, but thrust blocks or other restraints must be used to keep the joints from separating from the internal hydrostatic forces. When flanged valves are buried, the valve should be bolted to flanges of the adjacent piping with no other rigid support provided under or around the valve body. In no case should the valve body be supported directly by a saddle or other structure. In this condition, the valve becomes an anchor or a support for the piping system and will be forced to transmit piping loads.

Aboveground, exposed, or plant piping support design should carefully avoid, minimize, or not apply additional thrust or bending moments to the valve's end connections whenever possible. Buried applications should include either valve boxes or vaults that do not transmit piping, traffic, shock, settlement, or earth loads and stresses to either the valve or actuator as well as not apply additional thrust or bending moments to the valve's end connections.

If external and piping loads are severe enough, the valve may bind (from metal-to-metal interference) at various locations within the assembly including shafts, trunnions, seats, packings, and bearings, which may cause the valve to become inoperable.

Installation Concerns and Precautions

When quarter-turn valves have adjustable seating, install the seat-adjustment side of the valve for access and adjustment in service unless the manufacturer drawings or installation procedure indicate a different required orientation.

When practical, quarter-turn valves in buried installations should be located in vaults. Many types of buried pipes are designed to deflect 2–5 percent of pipe diameter, which is detrimental to the valve's integrity. Adjacent pipe should be supported and/or stiffened to provide a round mating connection that does not distort the valve body.

Foreign material in a quarter-turn valve can damage the seat when valves are operated. Be sure valve interiors and adjacent piping are cleaned of foreign material before making up valve-to-pipe joint connections. Foreign material may also prevent the valve from operating and damage the actuator.

Prepare pipe ends and install quarter-turn valves in accordance with the pipe manufacturer's instructions for the joint used. Do not deflect a pipe-valve joint. Do not use a quarter-turn valve as a jack to pull pipe into alignment. The installation procedure should minimize the bending of the quarter-turn valve with installation pipe loadings.

When quarter-turn valves are installed in vaults, the vault design should provide adequate space for removal or servicing of the valve-actuator assembly for purposes of repair and for access to adjust the thrust-bearing assembly. The vault should be designed and constructed wide enough such that the doors, covers, or panels on the actuator housing can be opened without the doors or panels striking an adjacent wall, thereby limiting or preventing access to the inside of the actuator housing or control system electronics. The possibility of groundwater or surface water entering the vault and the disposal of the water should be considered. The valve operating nut should be accessible from the top opening of the vault with a tee wrench.

As discussed previously, single-seated quarter-turn valves may have different shutoff characteristics based on seat orientation and should be installed with the seat oriented for shutoff in the direction preferred or required. Also, valves with adjustable, replaceable, or repairable seats may require access from a specified end of the valve to facilitate the rework. When this orientation, the installed piping arrangement, and the shutoff direction are in conflict, the designer may need to compensate, select a different valve, or select an orientation that is most preferred of the choices.

Quarter-turn valves in pump-control applications should be installed such that the seat orientation best prevents flow back toward the pump.

Buried quarter-turn valves should be installed with the shaft horizontal and the actuator input drive nut upward.

Valve box or extension pipe should be installed so that the actuator input drive nut and extension stem turn freely.

PVs in fluids free of suspended solids may be installed in any orientation relative to vertical unless there is an actuator orientation limitation. If practical, PVs shall be installed so that the pipeline pressure is exerting force on the plug from opposite the seat end of the valves (e.g., direct pressure orientation).

PVs used for fluids containing suspended solids should be installed as shown in Figure 6-14. When installed in horizontal pipes, the axis of the closure member shaft is to be horizontal, with flow entering the valve body from the seat end and keeping the plug in the upper half of the body when rotated (counterclockwise) to open. For pump isolation service with fluids containing suspended solids, the discharge valve is best installed with the seat downstream from the pump and the plug rotating into the top of the pipeline in the open position.

Quarter-turn valves in horizontal pipes used with fluids containing suspended solids are best installed with their shafts horizontal so that the sediment does not collect or penetrate into the shaft-bearing area and cause binding of the shaft. Even when used with clean fluids, one should avoid vertically oriented shafts unless there are other overriding conditions. Valves with horizontally oriented shafts tend to flush the bottom of the pipe for better long-term service.

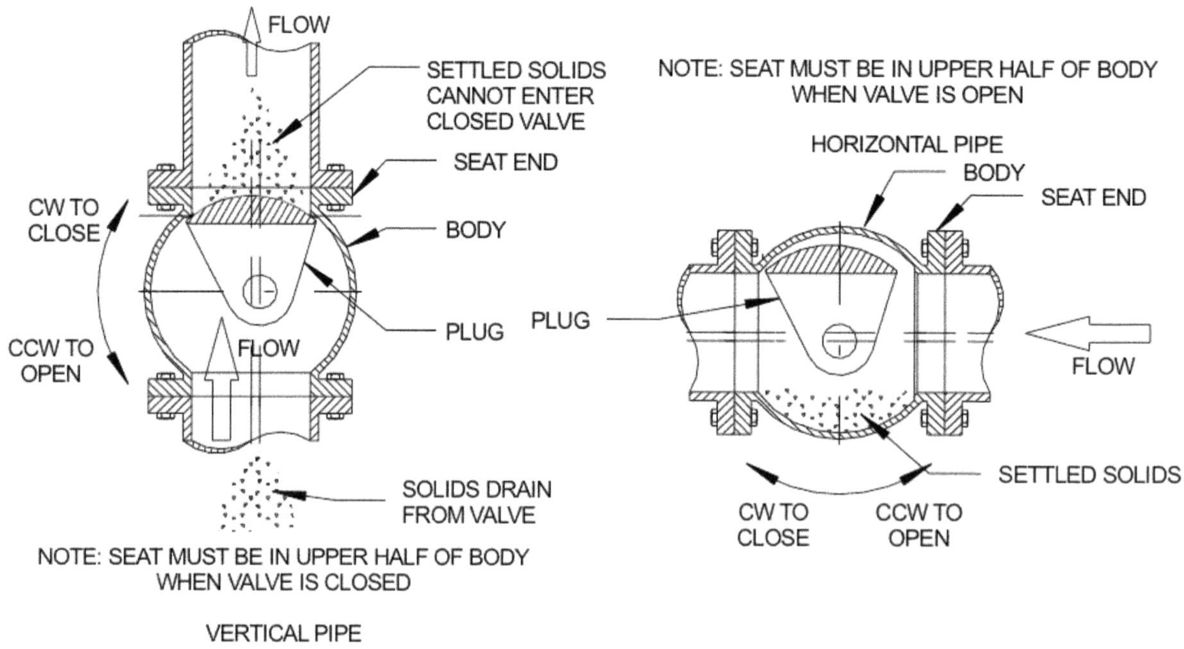

CCW: counterclockwise, CW: clockwise

Figure 6-14 PV preferred installation orientation for suspended solids applications

Whenever possible, the actuator should be mounted to the side or above a horizontal pipe. If mounting an actuator below the pipe is necessary, the valve and/or actuator manufacturer should be consulted regarding this orientation.

Double-seated valves may be susceptible to pressure locking between the seats. System designers should review pressure transients or excursions and operating procedures to assure that extreme pressure transients or pressure excursions are not captured between the seats or create seat binding, distortion, and high loads. Water works systems having operating thermal cycles with greater than a 50°F (28°C) temperature change (ΔT) are not typical. If this should be a system design condition, then the valve construction should be reviewed for both pressure locking and thermal binding caused by differences in material and fluid thermal coefficients of expansion that create higher friction-based torque components that could render the valve inoperable. This is of greatest concern for double-seated valves or with series-redundant valves operated simultaneously. Venting the internal central cavity of the valve or disabling one of the seals may be necessary. Such conditions should be reviewed with the valve manufacturer. Large or frequent thermal cycling can also cause grease to migrate past actuator housing seals. If this is a concern, vents, breathers, or compliant volumes may be needed on the actuators.

Loads from the Valve into the Pipe System

Water flowing through a partially open quarter-turn valve exerts drag and lift forces and dynamic torque on the valve's closure member. For a given valve, these fluid dynamic forces and moments vary with the valve position and flow through the valve. These are internal loads supported entirely within the valve's structure (e.g., body, closure member, shaft, and actuator). The forces and loads imposed on the pipe end connections are the same nozzle, differential pressure, and fluid dynamic loads that would be present with any valve type, orifice, or related piping component.

TYPICAL RANGE OF SOME COEFFICIENTS

As many users of this methodology may not have intimate knowledge of the various factors and coefficients involved, the following provides some direction regarding typical values found in the industry. These data were taken from a variety of sources and are not to be considered representative of any single valve or manufacturer. These are provided as informational only and should be understood as being only nominal representations.

An attempt to normalize was used, and outliers were removed as being not representative. All values shown are for valves and materials in good, well-maintained conditions and do not represent heavily degraded or misapplied conditions. These data are only provided for instructional and comparative use.

Bearing Friction

Table 6-1 provides some typical ranges for the bearing coefficient of friction, C_f, which may vary widely with the bearing and shaft materials. All valves within the scope of this report

Table 6-1 Typical bearing friction coefficients

Material Type	Low-Value C_f	Mean-Value C_f	High-Value C_f
Nonmetallic	0.07	0.12	0.25
Metallic	0.125	0.25	0.35

Table 6-2 Typical packing coefficients

Low-Value $C_{pck\ f}$	Mean-Value $C_{pck\ f}$	High-Value $C_{pck\ f}$
30 lb/in. (5,254 N/m)	400 lb/in. (70,000 N/m)	1,200 lb/in. (210,000 N/m)
Low-Value $C_{pck\ tq}$	Mean-Value $C_{pck\ tq}$	High-Value $C_{pck\ tq}$
100 lb (444 N)	450 lb (2,002 N)	700 lb (3,114 N)
Low-Value μ_P	Mean-Value μ_P	High-Value μ_P
0.10	0.20	0.30

Table 6-3 Typical seating coefficients

Valve Type	Low-Value C_{sc} *lb/in. (N/m)*	Mean-Value C_{sc} *lb/in. (N/m)*	High-Value C_{sc} *lb/in. (N/m)*
BV, resilient seats	10 (0.054)	24 (0.137)	40 (229)
BFV, resilient seats	6 (0.034)	16 (0.091)	36 (0.206)
Valve Type	Low-Value C_{sp} *lb/in./psi (N/m/kPa)*	Mean-Value C_{sp} *lb/in./psi (N/m/kPa)*	High-Value C_{sp} *lb/in./psi (N/m/kPa)*
BV, resilient seats	0.08 (2.03)	0.12 (3.05)	0.26 (6.60)
BFV, resilient seats	–0.02 (–0.51)	0.02 (0.51)	0.05 (1.27)

have iron, steel, or stainless-steel shafts, but the bearing material may be of many types. In general, nonmetallic bearings are designed to have a lower coefficient of friction than the metallic counterparts. Also, some nonmetallic bearing materials have coefficients of friction that vary with the applied load.

Packing Friction

Table 6-2 provides typical ranges for the packing coefficients, $C_{pck\ f}$ and $C_{pck\ tq}$, and coefficient of friction, μ_P, which may vary widely with the packing type and materials. Many packing manufacturers provide values and procedures to determine packing loads.

Seating Coefficients

Table 6-3 provides some typical ranges for the seating coefficients, C_{sc} and C_{sp}, which also vary widely by design and materials.

Flow Coefficients

Table 6-4 provides typical ranges for the flow coefficients, C_v and K_v, of quarter-turn valves. These coefficients can be expressed, grouped, and graphed in many ways. First, the full-open BFV flow resistance coefficient, K_v, is normally between 0.30 and 0.85. The C_v and K_v are inversely related, so the high K_v corresponds to the low C_v. Low pressure–rated valves generally have lower full-open flow resistance (K_v) values.

BFVs have inherent characteristic curves as given in Figure 2-4, which approach those of equal percentage control valves. These generalized curves can be used to calculate the C_v and K_v for BFVs at intermediate positions or angles between fully open and fully closed.

Table 6-4 Typical full-open flow coefficients

Valve Type	Low-Value C_v	Mean-Value C_v	High-Value C_v
BV and RCV	$118.9 \times D_{inches}^2$	$139.2 \times D_{inches}^2$	$188.8 \times D_{inches}^2$
BFV	$32.4 \times D_{inches}^2$	$41.0 \times D_{inches}^2$	$54.5 \times D_{inches}^2$
PV, red. port	$31.5 \times D_{inches}^2$	$33.5 \times D_{inches}^2$	$39.9 \times D_{inches}^2$
PV; full port	$38.9 \times D_{inches}^2$	$44.8 \times D_{inches}^2$	$57.4 \times D_{inches}^2$
Valve Type	High-Value K_v	Mean-Value K_v	Low-Value K_v
BV and RCV	0.063	0.046	0.025
BFV	0.85	0.53	0.30
PV, red. port	0.90	0.79	0.56
PV; full port	0.59	0.44	0.27

RCV: rotary cone valve, red.: reduced

Dynamic Torque Coefficients

Figure 3-16 provides typical ranges for BFV dynamic torque coefficients, C_t, which vary widely by design and materials. The combination of the flow and torque coefficients should be from a matched set of data. It should not be assumed that valves with high flow resistance values have low torque coefficient values. These data are generic and representative, but use of these data will not be representative of a specific valve.

CAUTIONS

Valves installed contrary to the recommendations in the previous section or in configurations subject to significant nonuniform or swirling upstream flow may develop torque requirements or stresses in excess of those generally assumed in sizing valve shafts, disc connections, and actuators. Consequently, electric actuators may stall, and other components may fail over time as a result of metal fatigue. Failure of any component (shaft or coupling) that connects the disc to the actuator mechanism may cause the disc to slam closed, with resulting damage to the valve and possibly severe water hammer and pipe damage.

Such an installation requires actuators with higher torque ratings and stronger valve components where improvements in upstream piping conditions are not feasible.

Actuator Speed of Operation

Fast valve opening and closure rates can impose large pressure transients in piping systems. The AWWA quarter-turn valve standards typically impose an operating time for power actuators of about 2–4 s per inch of nominal valve size. This is based on reasonable operating power requirements for the actuator and low transient pressures in a pipeline up to about 4,000 diameters in length. Operating speeds of operation should be reviewed by the user when pipe lengths are near or exceed this limit or when other operational requirements involve the use of faster speeds.

Actuator Servicing and Removal

The actuator should not be serviced or removed from installed valves unless the pipeline on both sides of the valve is de-energized (i.e., no pressure and drained). A quarter-turn valve closure member that is not restrained by the actuator may move, slam open, or slam closed, causing damage. Extreme caution must also be used when examining or working on a quarter-turn valve in the line. The actuator must be locked out to prevent unexpected travel. An installed valve without an actuator is a dangerous valve. An offset-style closure member may tend to open or close because of its offset and its closure member's center of gravity when installed. Similarly, hydrostatic force (discussed in Chapter 3) may also open an unsecured valve with a horizontal shaft in a horizontal line with water on one side of the closure member. Personnel should not enter a large-diameter pipe where any valve in the pipe has an unsecured actuator or a removed valve actuator. Personnel should not enter a large-diameter pipe where the actuator has not been locked and tagged out.

Seat Servicing and Removal

Valve seats should only be serviced or replaced with both sides of the valve de-energized (i.e., no pressure and drained). Many industrial safety jurisdictions or authorities consider valves to be pressure vessels that require the valve to be in a state where there is no pressure and no fluid on both sides when working on any pressure element or appurtenant component. Should valve seats be serviced or repaired inline, consideration must be given to confined space entry requirements of the governing municipality, jurisdiction, or safety authority—in particular, the regulations concerning single or double block and bleed isolation. When possible, removing the valve from the adjacent piping to repair or replace the valve seats is recommended.

Throttling Flow

Quarter-turn valves have good flow characteristics and are often used for throttling flow. However, quarter-turn valves are usually limited to a throttling range of 15–75 degrees open (15–85 percent). Operating valves at positions less than 15 degrees (15 percent) open may cause high localized velocities and cavitation, which can damage the seating surfaces.

As discussed in Chapter 4, cavitation can be observed by detecting a rumbling noise immediately downstream from the valve, similar to rocks flowing through the line, or by the use of an accelerometer attached to the pipe. Cavitation is a result of excessive pressure drop across the valve combined with low downstream pressure. When the localized pressure downstream of the disc falls below the vapor pressure of water (typically about 0.5 psia [3.5 kPa] for cold water), water vapor bubbles will form and then violently implode downstream as the pressure recovers (see Figure 4-1).

SUMMARY

The issues presented in this manual will assist users, system engineers, and designers to understand quarter-turn valve characteristics. The calculations, recommendations, applications, and valve installation precautions presented here will provide the user and designer with the most effective and trouble-free application of this type of valve. Because of the engineering complexity of this subject matter and the size dependency of the many variables, their individual effect increases with valve size. Greater effort should be placed on the details, methods, effects, and results as the valve size increases.

REFERENCES

ANSI (American National Standards Institute)/AWWA. 2021. C516. Large Diameter Rubber-Seated Butterfly Valves, 78 in. (2,000 mm) and Larger. Denver: AWWA.

ANSI/AWWA. 2021. C517. Resilient-Seated Cast-Iron Eccentric Plug Valves. Denver: AWWA.

ANSI/AWWA. 2021. C541. Hydraulic and Pneumatic Cylinder and Vane-Type Actuators for Valves and Slide Gates. Denver: AWWA.

ANSI/AWWA. 2021. C542. Electric Motor Actuators for Valves and Slide Gates. Denver: AWWA.

ANSI/AWWA. 2018. C507. Ball Valves, 6 in. Through 60 in. (150 mm through 1,500 mm). Denver: AWWA.

ANSI/AWWA. 2018. C519. High-Performance Waterworks Butterfly Valves—3 In. (75 mm) Through 60 In. (1,500 mm). Denver: AWWA.

ANSI/AWWA. 2015. C504. Rubber-Seated Butterfly Valves. Denver: AWWA.

This page intentionally blank.

Chapter 7

Control Valve Sizing and Selection

This chapter provides an aid for selection and sizing of valves for control valve service. The AWWA quarter-turn valves within the scope of this manual have inherent characteristics that are close to equal percentage, which make them good candidates for control service in mild service applications. These valves do not generally have added anticavitation or specialized control trim to extend their control range but can offer reasonable control service in some applications.

This chapter describes approaches to evaluate the applications and assist in the selection of a control valve. These methods are initial assessments for designing. More severe or critical service applications will require the use of valves specifically designed as control service valves.

The initial estimation is illustrated with typical valve designs, and once the final selection is made, the calculations should be performed with more precise data for the valve chosen.

VALVE TYPE SELECTION

Each quarter-turn valve in the AWWA standards has its design differences, advantages, and disadvantages. These include material selections, internal closure member (obturator) design, port shape, full or reduced port, tolerance for fluid medium additional content (debris or contaminants), or compatibility with fluid medium chemistry. This is part of selecting a control valve type or design regarding what is best for the service. These are covered in many other texts (such as the *Valve Handbook* by Philip L. Skousen). The valve types and construction are best selected by the system designer.

Although the same general principles covered in this chapter may apply to other valve types, the methods described here apply primarily to the four quarter-turn AWWA valves of this scope. If more than one AWWA valve type is otherwise acceptable for the system's purpose, then each valve type may be reviewed individually. As each valve's flow, cavitation, and choking characteristics are different, the result can vary between the valve types.

REQUIRED INPUTS

The following are inputs that should be obtained to permit good control valve selection.

Valve Types and Pressure Classes

Each of the AWWA valve types is available in at least one pressure class and fluid velocity class. If the system or system designer has any more stringent needs for considering any other pressure or flow limitations, these should be noted for exclusion purposes. Otherwise, the pressure class and flow designation of the applicable valves can be used as the limitations when determining the valve selections.

Installed Pipe Size

In any system, there are reasons for the selection of the basic pipe size in which the control valve is located. This size selection is one of the first inputs needed. A well-sized control valve is often smaller than the line size, and many valve sizes may be reviewed.

Pressures, Flow Rates, and Water Temperature

As the control valve's purpose is to dissipate energy (via pressure or flow control) before the fluid enters the downstream system, these temperatures, pressures, pressure losses, and fluid velocities or flow rates should be defined. There should be at least three separate, bounding, and specific operating conditions of upstream pressure, downstream pressure or differential pressure, fluid velocity or flow rate, and fluid (water) temperature. Often, there are four or more conditions covering a wide range of maximum and minimum upstream pressures, maximum and minimum downstream or differential pressures, maximum and minimum fluid velocities or flow rates, and maximum and minimum medium temperatures. The highest fluid temperature produces the greatest potential for cavitation and choking, but if the temperature range is wide, both extremes should be evaluated. If the fluid temperature range is narrow, a matrix of eight different control condition extremes (see Table 7-1) may be used at the high fluid temperature. The highest fluid temperature conditions are likely to be the most limiting in relation to choking and cavitation. The highest and lowest temperature conditions may be individually evaluated for sensitivity when the fluid temperature range is wide.

For pump-control applications, additional operating conditions may be selected from the multiple pump curves and the upper and lower bounding system conditions. Torque calculations for a pump curve (or a combination of pumps) can be performed, but they only represent the predicted operating torque for one set of conditions. Actuator sizing calculations should be performed at the highest closed differential pressure and the highest possible system flow rate (or line velocity).

Example Conditions to be Evaluated

This example will be based on a design of AWWA Class 150B butterfly valves (BFVs) and AWWA Class 150D ball valves (BVs) for a range of sizes below 30 in. (750 mm).

Table 7-1 Typical control valve sizing condition matrix

Case Reference No.	Upstream Pressure	Downstream Pressure	Differential Pressure	Flow Rate	Required Flow Coefficient CV or CVM
1	**Maximum**	**Maximum**	Moderate	**Maximum**	Moderate
2	**Maximum**	**Maximum**	Moderate	*Minimum*	Moderate
3	**Maximum**	*Minimum*	**Maximum**	**Maximum**	Moderate
4	**Maximum**	*Minimum*	**Maximum**	*Minimum*	*Minimum*
5	*Minimum*	**Maximum**	*Minimum*	**Maximum**	**Maximum**
6	*Minimum*	**Maximum**	*Minimum*	*Minimum*	Moderate
7	*Minimum*	*Minimum*	Moderate	**Maximum**	Moderate
8	*Minimum*	*Minimum*	Moderate	*Minimum*	Moderate

The assumed 6-in. (150-mm) through 30-in. (750-mm) BFVs have a mean coefficient factor, $C_{vfactor}$, equal to 40 times the diameter squared, and the 6-in. (150-mm) through 30-in. (750-mm) BVs assumed have a mean coefficient factor, $C_{vfactor}$, equal to C_v = 150 times the diameter squared.

At each of the identified flow-control conditions of upstream pressure, downstream or differential pressure, fluid flow rate, and specific gravity or fluid temperature, there is a required valve flow coefficient or a valve alone (in the case of a line-sized valve) for the valve, upstream reducer, and downstream expander.

The example conditions are for a 30-in. (750-mm) pipe flowing 60°F (16°C) water, an operating pressure upstream of the valve between 90 and 110 psig (620 and 758 kPa), an operating pressure downstream of the valve between 80 and 85 psig (552 and 596 kPa), an operating differential pressure between 5 and 30 psid (34.4 and 206 kPa-d), and an operating flow rate of 4,000 to 15,000 gpm (908 to 3,407 m^3/h). Table 7-2 displays the operating condition matrix and results of the required C_V calculations.

The following are the basic limits that start the process of control valve sizing. This method is based on the following equations from Chapter 2:

(In US customary units);

$$Q = C_v \times \sqrt{\frac{\Delta P}{S_g}} \quad \text{(Eq 2-1)}$$

(In SI metric units);

$$Q = C_{vm} \times \sqrt{\frac{\Delta P}{S_g}} \quad \text{(Eq 2-2)}$$

$$C_{vm} = C_v \times 0.86 \quad \text{(Eq 2-3)}$$

NOTES:

See Chapter 2 or the Nomenclature section for variable definitions and units of measure.

The AWWA standard sizes are US customary inches. The remainder of this chapter will be based primarily on US customary units for simplification. Conversions to metric units can be found in the Nomenclature section.

Table 7-2 Example operating conditions

Case Reference No.	Upstream Pressure psig (kPa)	Downstream Pressure psig (kPa)	Differential Pressure psid (kPa-d)	Flow Rate gpm (m^3/h)	Required Valve and Fittings Flow Coefficient C_v (C_{vm})
Input	Input	Input	Calc.	Input	Calc.
1	110 (758)	85 (596)	25 (172)	15,000 (3,407)	3,000 (2,580)
2	110 (758)	85 (596)	25 (172)	4,000 (908)	800 (688)
3	110 (758)	80 (52)	30 (207)	15,000 (3,407)	2,739 (2,356)
4	110 (758)	80 (52)	30 (207)	4,000 (908)	730 (627)
5	90 (620)	85 (596)	5 (34)	15,000 (3,407)	6,708 (5,769)
6	90 (620)	85 (596)	5 (34)	4,000 (908)	1,789 (1,539)
7	90 (620)	80 (52)	10 (69)	15,000 (3,407)	4,743 (4,079)
8	90 (620)	80 (52)	10 (69)	4,000 (908)	1,265 (1,088)

Calc.: calculation

LIMITATION CALCULATIONS

Largest Valve Size to be Evaluated

The first limitation is that the selected size should not exceed the pipe size. This is a basic and required input and should be defined.

Smallest Velocity-Based Valve Size to be Evaluated

For the valve type and pressure class designation, there is a rated maximum velocity or flow rate designation (V_{class}). These are typically velocities of 8, 16, 24, or 35 feet per second (2.44, 4.88, 7.32, or 10.67 meters per second) for many AWWA valves. As the typical US customary units of measure for these calculations are flow in gallons per minute, velocity in feet per second, and valve size in inches, the smallest possible valve size is calculated as:

$$D_{min-size} = \sqrt{\frac{0.4085 \times Q_{max-gpm}}{V_{class}}} \qquad \text{(Eq 7-1)}$$

At this point, the largest and the smallest should be in the correct order and should define a range of several sizes. If this does not result in a range of sizes, a different valve or a greater rated maximum velocity or flow rate (V_{class}) should be considered.

For example BFV, this would be $(0.4085 \times 15{,}000/16)^{1/2}$ or 19.6 in. rounded up to 20 in. (500 mm). For the example BV, this would be $(0.4085 \times 15{,}000/35)^{1/2}$ or 13.2 in. rounded up to 14 in. (350 mm).

Grouping of Valve Flow Coefficients, C_V

After the valve type and design have been selected, the general characteristics of the series should be determined. From the valve's published or manufacturer's data for the selected valve designs in sizes equal to the pipe size and several sizes smaller, develop a mean $C_{vfactor}$ such that the group of coefficients is representative of this $C_{vfactor}$ times the valve diameter squared ($C_{vfactor} = C_v/\text{diameter}^2$).

Inherent Flow Characteristics

The inherent flow characteristics for the valve's published or manufacturer's data for the valve designs selected should be used. This example is based on the typical inherent BV or rotary cone valve (RCV) and typical inherent BFV characteristics as provided in Figure 2-4.

Assuming the typical inherent BV or RCV and BFV characteristics of Figure 2-4, this chapter is based on the percentage of full-open C_v at 15 degrees open and 75 degrees open as provided in Table 7-3. The fully open C_v of the valves is based on the full-open C_v flow coefficient of either 40 (for the BFV) or 150 (for the BV) times the diameter squared as described in the previous section. A summary by size is provided in Table 7-4.

The greatest and the smallest required C_v defines the valve's full control range. In Table 7-1, these are the case reference numbers 4 and 5.

The valve control positions should not be right at the fully open position or too close to the fully closed position. This allows for some consideration of design changes or errors at the most fully open control position and reduces erosion and extremely high localized fluid velocities at the most closed control position. The greatest required valve C_v (case reference number 5) should occur at a valve angle less than 75 degrees (83 percent) open, and the lowest required C_v (case reference number 4) should occur at a valve angle greater than 15 degrees (17 percent) open.

Table 7-3 Typical percentage of full-open C_v (or C_{VM}) at 15 and 75 degrees open

Angle	Ball or Cone Valve Typical % Full-Open C_v or C_{vm}	Butterfly Valve Typical % Full-Open C_v or C_{vm}
15	1.30%	2.78%
75	28.61%	75.40%

Table 7-4 Some typical AWWA valve flow coefficients for control valve sizing

Valve Size	Ball Valve 90° C_v (C_{vm})	Ball Valve 75° C_v (C_{vm})	Ball Valve 15° C_v (C_{vm})	Butterfly Valve 90° C_v (C_{vm})	Butterfly Valve 75° C_v (C_{vm})	Butterfly Valve 15° C_v (C_{vm})
6	5,400 (4,644)	1,545 (1,329)	70 (60)	1,440 (1,238)	1,086 (934)	40 (34)
8	9,600 (8,256)	2,746 (2,362)	125 (108)	2,560 (2,202)	1,930 (1,660)	71 (61)
10	15,000 (12,900)	4,291 (3,690)	196 (169)	4,000 (3,440)	3,016 (2,594)	111 (95)
12	21,600 (18,576)	6,179 (5,314)	282 (243)	5,760 (4,954)	4,343 (3,735)	160 (138)
14	29,400 (25,284)	8,411 (7,233)	383 (329)	7,840 (6,742)	5,912 (5,084)	218 (187)

(continued)

Table 7-4 Some typical AWWA valve flow coefficients for control valve sizing *(Continued)*

Valve Size	Ball Valve 90° C_v (C_{vm})	Ball Valve 75° C_v (C_{vm})	Ball Valve 15° C_v (C_{vm})	Butterfly Valve 90° C_v (C_{vm})	Butterfly Valve 75° C_v (C_{vm})	Butterfly Valve 15° C_v (C_{vm})
16	38,400 (33,024)	10,986 (9,448)	501 (431)	10,240 (8,806)	7,721 (6,640)	285 (245)
18	48,600 (41,796)	13,904 (11,957)	634 (545)	12,960 (11,146)	9,772 (8,404)	361 (310)
20	60,000 (51,600)	17,165 (14,762)	782 (673)	16,000 (13,760)	12,064 (10,375)	445 (383)
24	86,400 (74,304)	24,718 (21,257)	1,127 (969)	23,040 (19,814)	17,373 (14,941)	641 (551)
30	135,000 (116,100)	38,621 (33,214)	1,760 (1,514)	36,000 (30,960)	27,145 (23,345)	1,002 (862)

Smallest Head Loss–Based Valve Size to be Evaluated

From the example maximum required C_v of 6,708 (C_{vm} = 5,769), the smallest possible head loss–based valve sizes can be chosen by finding the valves that have a 75-degree open C_v greater than the maximum required. For valve sizes below the pipe size, the combined C_v of the valve, reducer, and expander will be less than that of the valve alone. Therefore, some allowance for this C_v reduction should be considered.

The BVs greater than 14 in. (350 mm) and BFVs greater than 16 in. (400 mm) have a 75-degree C_v greater than the maximum required of 6,708 (C_{vm} = 5,769) calculated in Table 7-2. As these sizes are substantially smaller than the pipe size, the range for consideration will be limited to larger valve sizes. For this example, BVs less than 16 in. (400 mm) and BFVs less than 18 in. (450 mm) may be reasonably eliminated from consideration.

Largest Head Loss–Based Valve Size to be Evaluated

The minimum required C_v of the example is 730 (C_{vm} = 628). For the largest valve size elimination, find the largest valve sizes that have a 15-degree C_v less than 730 (C_{vm} = 628). For this example, BVs exceeding 20 in. (500 mm) and BFVs exceeding 24 in. (600 mm) may be reasonably eliminated from consideration.

Suggested Evaluation Size Ranges:

The BV minimum size based on the maximum velocity is 14 in., and the minimum size based on the head loss is 16 in. (400 mm). The minimum BV size suggested is 16 in. (400 mm). The largest BV is either the pipe size of 30 in. or the maximum head-loss size of 20 in. (500 mm). The maximum BV size suggested is then 20 in. (500 mm). The suggested size range for a BV is then 16–20 in. (400–500 mm). A BV diameter of 18 in. (450 mm) or 20 in. (500 mm) would be the initial size for evaluation.

The BFV minimum size based on the maximum velocity is 20 in. (500 mm), and the minimum size based on the head loss is 18 in. (450 mm). The minimum BFV size suggested is 20 in. (500 mm). The largest BFV is either the pipe size of 30 in. (750 mm) or the maximum head-loss size of 24 in. (600 mm). The suggested size range for a BFV is then 20–24 in. (500–600 mm). A BFV diameter of 20 in. (500 mm) or 24 in. (600 mm) would be the initial size for evaluation.

Control Valve Assembly C_v (or C_{vm}) Coefficients and Inherent Characteristic

Unless the indicated valve size is the same as the pipe size, the flow coefficients for this completed assembly (reducer, valve, and increaser) and the assembly's inherent characteristic curve are not the same as those of the valve alone. Use the Chapter 2 Reducer Installations equations to calculate the assembly C_v (or C_{vm}) for each valve angle and construct the assembly's inherent C_v (or C_{vm}) curve. Refer to Tables 2-1 and 2-2 for an example calculation. Once the assembly flow coefficients by valve position are developed, the required valve position for each of case references condition can be selected. Interpolation or curve-fit techniques can be used.

Figure 7-1 depicts the inherent characteristic curves for a 20-in. (500-mm) BV alone and the assembled 30-in. (750-mm) × 20-in. (500-mm) reducer, a 20-in. (500-mm) BV, and a 20-in. (500-mm) × 30-in. (750-mm) increaser, as well as the 20-in. (500-mm) BFV alone and the assembled 30-in. (750-mm) × 20-in. (500-mm) reducer, a 20-in. (500-mm) BFV, and a 20-in. (500-mm) × 30-in. (750-mm) increaser.

Differential Pressure and Flow Rate Operating Window Development

Each of the conditions to be evaluated in Table 7-1 are singular conditions, but more often, these extremes are not the normal operating conditions. One method of review with a clear picture is to use operating windows of differential pressure, flow rate, and cavitation indexes and valve position. These windows are graphs of differential pressure, flow rate, and the operating cavitation index versus the valve position required for each case reference condition. On each of the graphs, a window drawn to encircle all the case results will form the operating window.

Figures 7-2 and 7-3 display the 20-in. (500-mm) and 18-in. (450-mm) BV windows for the example conditions. Inspection of these windows shows that the 20-in. valve stays between 15 and 75 degrees open and a slightly larger window. Therefore, the 20-in. (500-mm) BV is suggested.

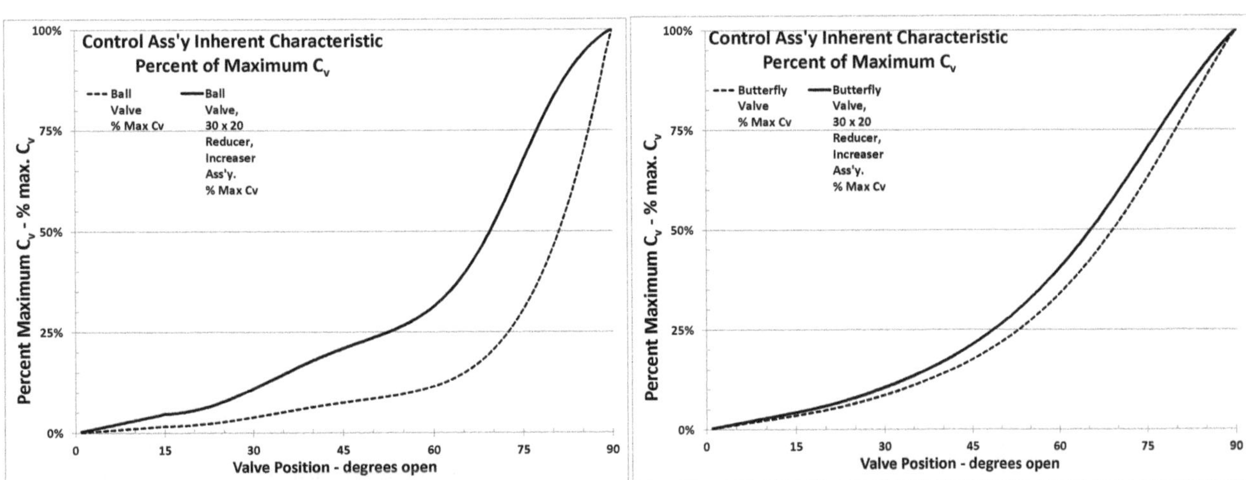

Figure 7-1 Control valve assembly inherent characteristic curves for a 20-in. (500-mm) BV alone and the assembled 30-in. (750-mm) × 20-in. (500-mm) reducer, a 20-in. (500-mm) BV, and a 20-in. (500-mm) × 30-in. (750-mm) increaser

NOTE: The percentage of maximum C_v or C_{vm} is the same. Control Ass'y: control valve assembly

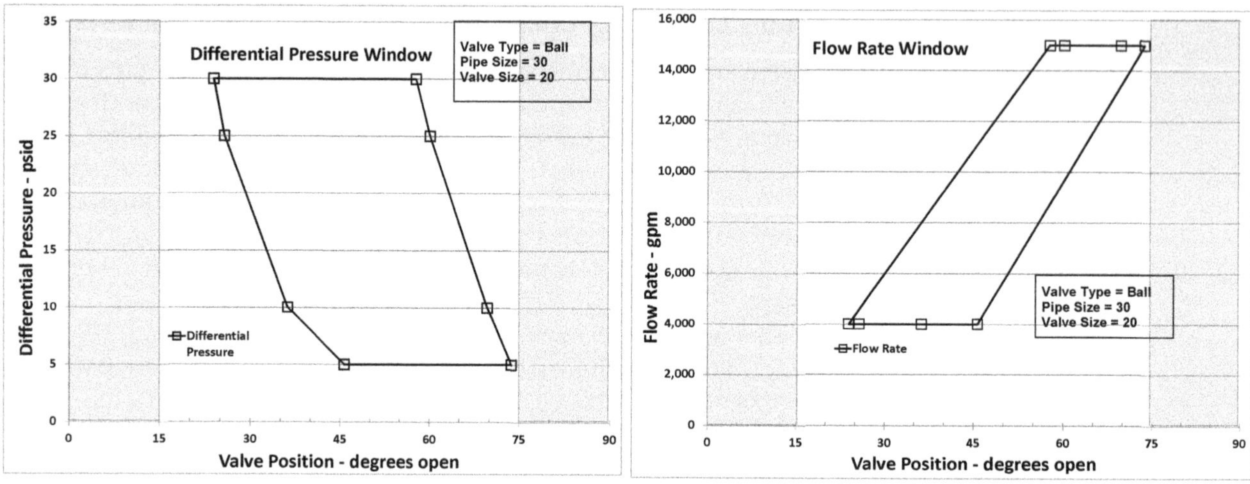

Figure 7-2 Example 20-in. (500-mm) BV differential pressure and flow rate windows

Note: Graphs in metric units appear the same. Only the vertical axis scale changes. Metric differential pressure is 34.5 kPa per division. Metric flow is 454 m^3/h per division.

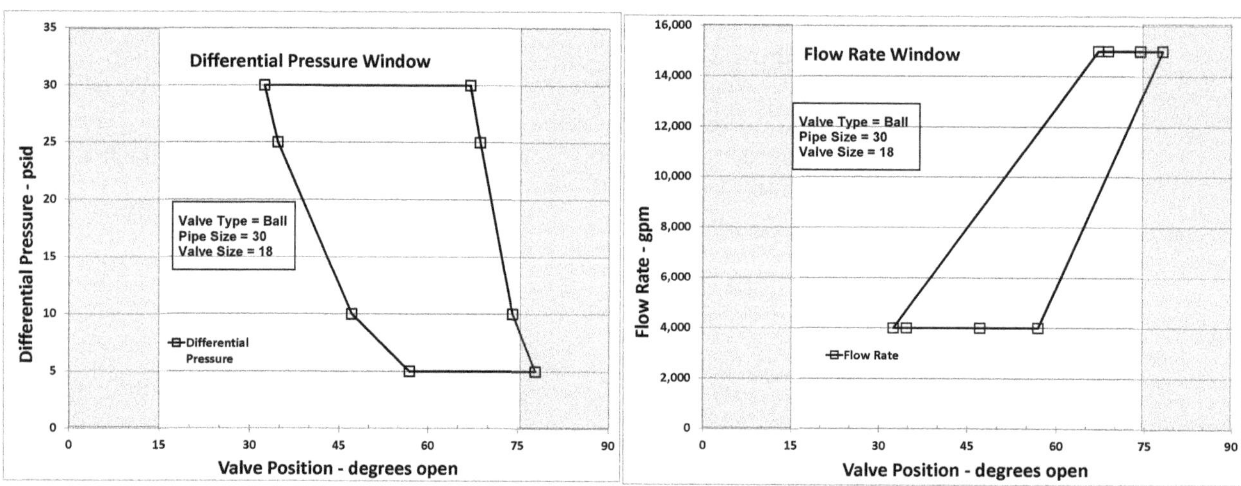

Figure 7-3 Example 18-in. (450-mm) BV differential pressure and flow rate windows

Note: Graphs in metric units appear the same. Only the vertical axis scale changes. Metric differential pressure is 34.5 kPa per division. Metric flow is 454 m^3/h per division.

Figures 7-4 and 7-5 display the 24-in. (600-mm) and 20-in. (500-mm) BFV windows for the example conditions. Examination of these windows shows that the 20-in. (500-mm) valve stays between 15 and 75 degrees open and a larger window. Therefore, the 20-in. (500-mm) BFV is indicated.

For this example, either a 20-in. (500-mm) BV or a 20-in. (500-mm) BFV is indicated as the most appropriate valve size selection.

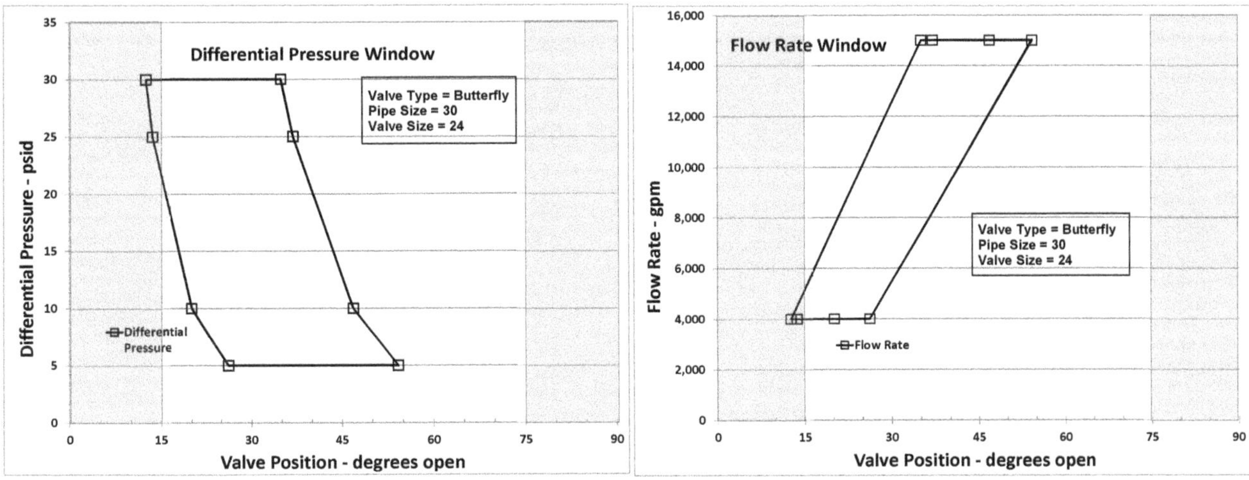

Figure 7-4 Example 24-in. (600-mm) BFV differential pressure and flow rate windows

Note: Graphs in metric units appear the same. Only the vertical axis scale changes. Metric differential pressure is 34.5 kPa per division. Metric flow is 454 m^3/h per division.

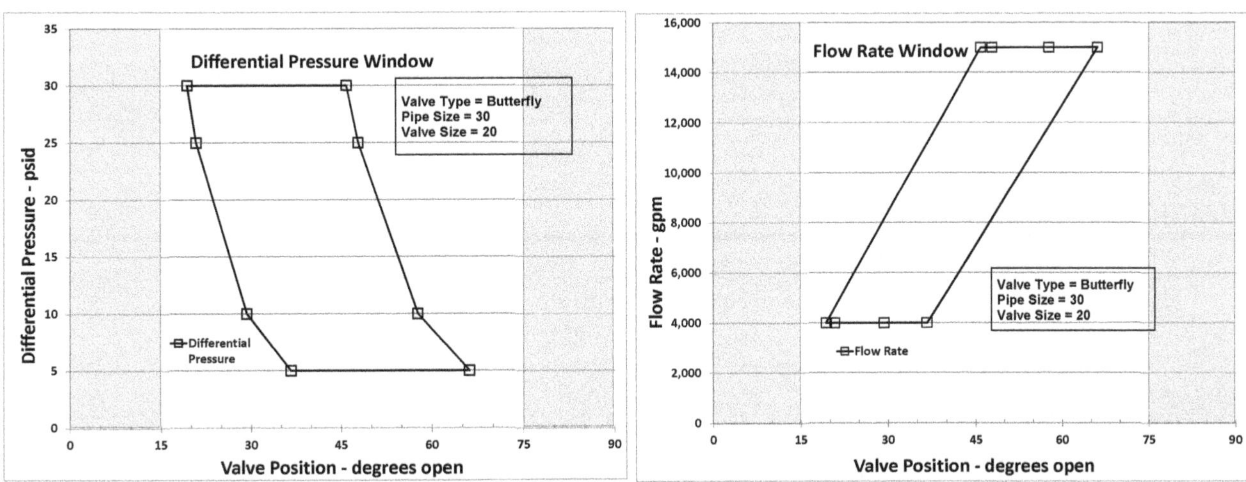

Figure 7-5 Example 20-in. (500-mm) BFV differential pressure and flow rate windows

Note: Graphs in metric units appear the same. Only the vertical axis scale changes. Metric differential pressure is 34.5 kPa per division. Metric flow is 454 m^3/h per division.

Choking and Cavitation Window Development

The valve's cavitation indexes and the choking coefficients were developed for straight line pipe flow and are not truly indicative of a reduced valve size control assembly. However, these data can be used as representative. Fully modeling the system and performing multiple condition calculations after valve selection will provide greater assurance of in-service performance where a closer review is considered necessary.

For each case reference condition, the valve angle, pressure upstream of the valve, and the choked differential pressure (ΔP_{choked}) can be calculated using Eq 2-21. If the ΔP_{choked} is less than the required ΔP, then choking is probable and should be evaluated further. Tables 7-5 and 7-6 show that the minimum downstream pressure of 80 psig (52 kPa) for this example is sufficient to suppress choking at the valves for all the case reference number conditions.

Table 7-5 BFV choking estimates

Operating Conditions					Butterfly Valve Estimates			
Ref. Case	Pup psig (*kPa*)	ΔP psid (*kPa-d*)	Flow Rate gpm (m^3/h)	Req'd. Valve C_v (C_{vm})	Valve Angle degrees open	Liquid Pressure Recovery Factor $F_{l\theta}^2$	ΔP_{choked} Eq. 2-21 psid (*kPa-d*)	Choked *Yes/No*
1	110 (758)	25 (172)	15,000 (3,407)	3,000 (2,580)	48	0.59	73.69 (508)	No
2	110 (758)	25 (172)	4,000 (908)	800 (688)	21	0.67	83.44 (575)	No
3	110 (758)	30 (207)	15,000 (3,407)	2,739 (2,356)	46	0.60	74.62 (514)	No
4	110 (758)	30 (207)	4,000 (908)	730 (627)	19	0.67	83.76 (578)	No
5	90 (620)	5 (34)	15,000 (3,407)	6,708 (5,769)	66	0.51	52.97 (365)	No
6	90 (620)	5 (34)	4,000 (908)	1,789 (1,539)	37	0.63	65.95 (455)	No
7	90 (620)	10 (69)	15,000 (3,407)	4,743 (4,079)	58	0.55	57.40 (396)	No
8	90 (620)	10 (69)	4,000 (908)	1,265 (1,088)	29	0.65	68.09 (469)	No

Ref.: reference, Req'd.: required

Table 7-6 BV choking estimates

Operating Conditions					Ball Valve Estimates			
Ref. Case	Pup psig (*kPa*)	ΔP psid (*kPa-d*)	Flow Rate gpm (m^3/h)	Req'd. Valve C^V (C_{vm})	Valve Angle degrees open	Liquid Pressure Recovery Factor $F_{l\theta}^2$	ΔP_{choked} Eq. 2-21 psid (*kPa-d*)	Choked *Yes/No*
1	110 (758)	25 (172)	15,000 (3,407)	3,000 (2,580)	60	0.66	82.29 (567)	No
2	110 (758)	25 (172)	4,000 (908)	800 (688)	26	0.79	98.35 (678)	No
3	110 (758)	30 (207)	15,000 (3,407)	2,739 (2,356)	58	0.68	84.77 (584)	No
4	110 (758)	30 (207)	4,000 (908)	730 (627)	24	0.79	98.36 (678)	No
5	90 (620)	5 (34)	15,000 (3,407)	6,708 (5,769)	74	0.50	52.67 (363)	No
6	90 (620)	5 (34)	4,000 (908)	1,789 (1,539)	46	0.75	78.58 (542)	No
7	90 (620)	10 (69)	15,000 (3,407)	4,743 (4,079)	70	0.56	58.43 (403)	No
8	90 (620)	10 (69)	4,000 (908)	1,265 (1,088)	36	0.78	81.49 (562)	No

Ref.: reference, Req'd.: required

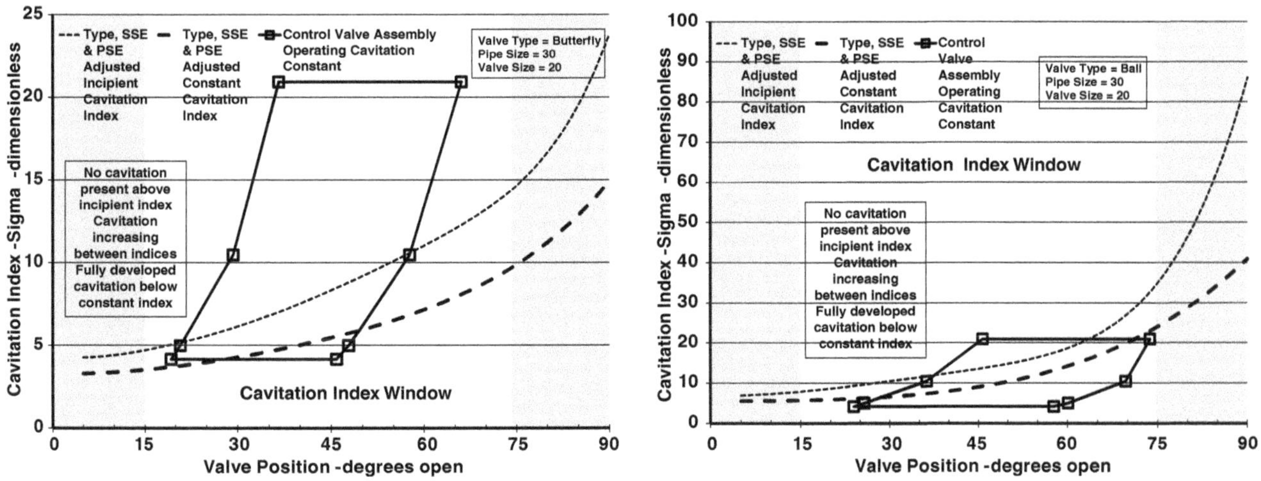

PSE: pressure scale effect, SSE: size scale effect

Figure 7-6 Example 20-in. (500-mm) BFV and 20-in. (500-mm) BV operating cavitation windows

Finally, the operating cavitation constant can be calculated. The adjusted incipient and constant indexes shown in Figure 7-6 are relative to the valve as inline tested. These may be the only available data for comparison. For this window, the major concern is the portion of the window that is below the constant cavitation line. For this example, the use of a 20-in. (500-mm) BFV is indicated and worthy of further investigation, as only a small portion of the operating window suggests constant cavitation is present.

SUMMARY

The control valve sizing methods and the development of the differential pressure, flow rate, and cavitation operating windows of this chapter can greatly assist in the selection of a control valve type and size. If a good selection is not found, finding a control valve specifically designed for more severe service may be necessary.

REFERENCES

ANSI (American National Standards Institute)/AWWA. 2021. C516. Large Diameter Rubber-Seated Butterfly Valves, 78 in. (2,000 mm) and Larger. Denver: AWWA.

ANSI/AWWA. 2021. C517. Resilient-Seated Cast-Iron Eccentric Plug Valves. Denver: AWWA.

ANSI/AWWA. 2021. C541. Hydraulic and Pneumatic Cylinder and Vane-Type Actuators for Valves and Slide Gates. Denver: AWWA.

ANSI/AWWA. 2021. C542. Electric Motor Actuators for Valves and Slide Gates. Denver: AWWA.

ANSI/AWWA. 2018. C507. Ball Valves, 6 in. Through 60 in. (150 mm through 1,500 mm). Denver: AWWA.

ANSI/AWWA. 2015. C504. Rubber-Seated Butterfly Valves. Denver: AWWA.

Skousen, P.L. 2011 (3rd ed.). *Valve Handbook*. New York: McGraw-Hill.

This page intentionally blank.

APPENDIX A

System

1. Owner/Client: ____________________
2. Project/Site: ____________________
3. Service/system description: ____________
4. Equipment ID or Tag No.(s): ____________
5. Type of installation: (select one)
   ❑ buried, ❑ submerged, ❑ nonburied/ submerged
6. Installed Location Type: (select one)
   ❑ indoors, ❑ outdoors, ❑ vault,
   ❑ other/data: ____________________
7. System Fluid: (select one)
   ❑ water, ❑ wastewater, ❑ gas ____________
   ❑ other/data: ____________________
8. Max. nonshock line pressure: ____________
9. Max. nonshock closed differential pressure:____
10. If known, maximum transient pressure; rate; characteristics: ____________________
11. Port fluid flow or velocity through valve & units:
    a. Normal conditions: ____________
    b. Maximum flow conditions: ____________
    c. Emergency opening: ____________
    d. Emergency closing: ____________
12. Fluid temperature range: ____________
13. Flow Direction: (select a + i, a + ii, a + iii, b, or c.)
    a. ❑ Uni-directional flow
       I. ❑ Manufactures best or preferred (TBD)
       II. ❑ Direct pressure/shaft side flow
       III. ❑ Reverse pressure/seat side flow
    b. ❑ Bidirectional flow
    c. ❑ Pump check operation
14. Service: (select one)
    ❑ On/off, ❑ Modulating, ❑ Throttling (set & held)
    For modulating or throttling provide (with units):
    a. Condition 1: Flow; ____________
       Upstream P; ____________
       Downstream P;____________
    b. Condition 2: Flow; ____________
       Upstream P; ____________
       Downstream P;____________
    c. Condition 3: Flow; ____________
       Upstream P; ____________
       Downstream P;____________
    d. Condition 4: Flow; ____________
       Upstream P; ____________
       Downstream P;____________

Installation (if known)

1. Description of connecting piping:
   ____
2. Installed pipe orientation: ❑ vertical, ❑ horizontal,
   ❑ other;____________________
3. Installed shaft orientation: ❑ vertical, ❑ horizontal,
   ❑ other; ____________________

Valve

1. Type: (select one) ❑ Ball (BV), ❑ Butterfly (BFV), ❑ Cone (RCV), ❑ Plug (PV)
2. Size of the valve & units: ____________
3. Press. class/vel. designation (AWWA): ________
4. Quantity required: ____________
5. Type of body end connection: (select one)
   ❑ Flanged, ❑ Wafer, ❑ MJ,
   ❑ Other: ____________________
6. Special materials or material requirements
   a. Body:____________________
      Closure member: ____________
      Metal seat(s): ____________
      Resilient seat: ____________
      Bearing: ____________
      Shaft: ____________
      Shaft seals: ____
7. If BV or RCV, ❑ double-seated, or ❑ single-seated

Actuator Mounting

1. Type: (select one)
   ❑ Direct, ❑ Replaceable packing yoke or bonnet, ❑ Extended actuator yoke or bonnet; C/L valve to C/L actuator input; ____________
   ❑ Extended actuator floor-stand & torque-tube; C/L valve to floor; ________________________
   ❑ Other: ________________________________

❑ Manual Actuator

1. Type: (select one or more) ❑ Direct (lever/nut), ❑ Worm gear, ❑ Traveling nut, ❑ Other
2. Input: (select one or more)
   ❑ Lever, ❑ Crank, ❑ 2″ nut, ❑ Chain lever, ❑ Handwheel, ❑ Chainwheel
3. Opening direction: (select one)
   ❑ Std.-CCW (Open Lft), ❑ Non-Std.-CW (Open Rt)
4. Input accessories/devices: (select one or more)
   ❑ Extension shaft, length; ________________
   ❑ Floor stand,
   ❑ Special handwheel diameter, ____________
   ❑ Special position indicator: _______________
5. Output accessories/devices: (select one or more)
   ❑ Limit switches, qty./type: _______________
   ❑ Position transmitter: ___________________

❑ Power Actuator

Attach separate data sheet. See AWWA C541 or C542.

Coatings & Other Special Requirements

____

____

____

____

____

APPENDIX B

Example of First-Principles Derivation of Seating Torque for Circular Seat

Let:

C_s	=	Seating friction load, force / unit length (i.e., pound–force / inch or Newtons / meter)
D	=	Diameter of the line of closure member seat contact = 2 × *R*, unit length (i.e., inches or meters)
L	=	Effective lever length = $R \times sin\theta$, unit length (i.e., inches or meters)
R	=	Radius of the line of disc seat contact, unit length (i.e., inches or meters)
T_{seat}	=	Seating torque, force–unit length (i.e. pound-force–inches or Newton-meters)
W	=	Friction load at seat contact circle, unit force = $Cs \times R \times \delta\theta$
Θ	=	Angle about the seat contact circle, radians

$$T_{seat} = 2 \times \int_0^{\pi} L \times W$$

$$T_{seat} = 2 \times \int_0^{\pi} (R \sin\theta) \times (C_S \times R \times d\theta)$$

$$T_{seat} = 2 \times C_s \times R^2 \int_0^{\pi} \sin\theta \, d\theta$$

$$T_{seat} = 2 \times C_s \times R^2 \times \left[-\cos\theta\right]_0^{\pi}$$

$$T_{seat} = 2 \times C_s \times R^2 \times \left[(1) - (-1)\right]$$

$$T_{seat} = 4 \times C_s \times R^2$$

$$T_{seat} = C_s \times D^2$$

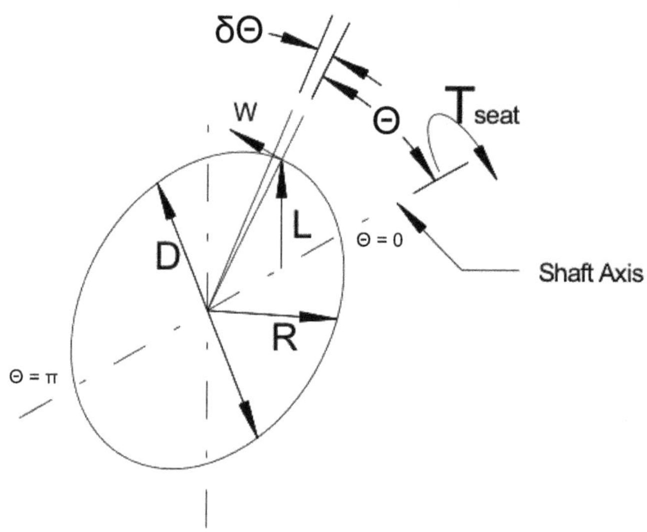

Figure B-1 Seating Torque Free-Body Diagram

Index

Note: *f* indicates figure; *t* indicates table

F

G

H

I

L

M

N

O

P

Q

R

S

This page intentionally blank.

AWWA Manuals

M1, *Principles of Water Rates, Fees, and Charges*, #30001
M2, *Instrumentation and Control*, #30002
M3, *Safety Management for Water Utilities*, #30003
M4, *Water Fluoridation Principles and Practices*, #30004
M5, *Water Utility Management*, #30005
M6, *Water Meters—Selection, Installation, Testing, and Maintenance*, #30006
M7, *Problem Organisms in Water: Identification and Treatment*, #30007
M9, *Concrete Pressure Pipe*, #30009
M11, *Steel Pipe—A Guide for Design and Installation*, #30011
M12, *Simplified Procedures for Water Examination*, #30012
M14, *Backflow Prevention and Cross-Connection Control: Recommended Practices*, #30014
M17, *Fire Hydrants: Installation, Field Testing, and Maintenance*, #30017
M19, *Emergency Planning for Water and Wastewater Utilities*, #30019
M20, *Water Chlorination/Chloramination Practices and Principles*, #30020
M21, *Groundwater*, #30021
M22, *Sizing Water Service Lines and Meters*, #30022
M23, *PVC Pipe—Design and Installation*, #30023
M24, *Planning for the Distribution of Reclaimed Water*, #30024
M25, *Flexible-Membrane Covers and Linings for Potable-Water Reservoirs*, #30025
M27, *External Corrosion Control for Infrastructure Sustainability*, #30027
M28, *Rehabilitation of Water Mains*, #30028
M29, *Water Utility Capital Financing*, #30029
M30, *Precoat Filtration*, #30030
M31, *Distribution System Requirements for Fire Protection*, #30031
M32, *Computer Modeling of Water Distribution Systems*, #30032
M33, *Flowmeters in Water Supply*, #30033
M36, *Water Audits and Loss Control Programs*, #30036
M37, *Operational Control of Coagulation and Filtration Processes*, #30037
M38, *Electrodialysis and Electrodialysis Reversal*, #30038
M41, *Ductile-Iron Pipe and Fittings*, #30041
M42, *Steel Water-Storage Tanks*, #30042
M44, *Distribution Valves: Selection, Installation, Field Testing, and Maintenance*, #30044
M45, *Fiberglass Pipe Design*, #30045
M46, *Reverse Osmosis and Nanofiltration*, #30046
M47, *Capital Project Delivery*, #30047
M48, *Waterborne Pathogens*, #30048
M49, *Quarter-Turn Valves: Head Loss, Torque, and Cavitation Analysis*, #30049
M50, *Water Resources Planning*, #30050
M51, *Air Valves: Air-Release, Air/Vacuum and Combination*, #30051
M52, *Water Conservation Programs—A Planning Manual*, #30052
M53, *Microfiltration and Ultrafiltration Membranes for Drinking Water*, #30053
M54, *Developing Rates for Small Systems*, #30054
M55, *PE Pipe—Design and Installation*, #30055
M56, *Nitrification Prevention and Control in Drinking Water*, #30056
M57, *Algae: Source to Treatment*, #30057
M58, *Internal Corrosion Control in Water Distribution Systems*, #30058
M60, *Drought Preparedness and Response*, #30060
M61, *Desalination of Seawater*, #30061
M62, *Membrane Applications for Water Reuse*, #30062
M63, *Aquifer Storage and Recovery*, #30063
M65, *On-Site Generation of Hypochlorite*, #30065

M66, *Cylinder and Vane Actuators and Controls—Design and Installation,* #30066

M68, *Water Quality in Distribution Systems,* #30068

M69, *Inland Desalination and Concentrate Management,* #30069

M77, *Condition Assessment of Water Mains,* #30077